Sea Energy Agriculture

by Maynard Murray, M.D.
with Tom Valentine

Sea Energy Agriculture

by Maynard Murray, M.D.

with Tom Valentine

Revised Second Edition

Acres U.S.A.
Austin, Texas

Sea Energy Agriculture

Acres U.S.A.
P. O. Box 91299
Austin, Texas 78709 U.S.A.
(512) 892-4400   fax (512) 892-4448
info@acresusa.com   www.acresusa.com

Printed in the United States of America

Publisher's Cataloging-in-Publication

Murray, Maynard, 1910-1983
Sea Energy Agriculture/Maynard Murray with Tom Valentine,
Austin, TX: Acres U.S.A. 2003.
viii, 109 pp., tables, photos.

2nd edition, revised.
Includes index.
Originally published by Valentine Books, Winston-Salem, North Carolina.

ISBN: 0-911311-70-X

1. Trace elements in agriculture. 2. Trace elements in plant nutrition.
3. Trace elements in animal nutrition. 4. Micronutrient fertilizers. I. Murray,
Maynard, 1910-1983. II. Valentine, Tom, 1935- III. Title.

S587.5.T7   631.81

Contents

Acknowledgment

This book contains a report of many years of research on sea salt agricultural technology conducted by the author. The opinions, assumptions and conclusions stated within the context of this work represent results of preliminary research conducted to date. Therefore, the material in this book is not intended to be accepted as conclusive in regard to methodology or results achieved and in no way should be construed to make a claim for better health or concrete advances in disease prevention and/or treatment. This report is presented in the interest of science to stimulate further research which will one day render conclusive results, hopefully for the benefit of all mankind.

A Note From Acres U.S.A.

"Life is electrical." With that statement Maynard Murray, M.D., opened his lecture at the 1976 Acres U.S.A. Conference in Kansas City. His many years of research then were ready to be put on the table. "There can be no life without a transfer of electrical energy."

This was a profound statement for an audience that had come to hear about organiculture and hadn't considered the business of conductivity and electricity, especially as it related to crop production. It didn't take Murray long to explain that the center of life's gravity was the ocean, the repository of minerals from the land, dissolved and carried to nature's settling basin via streams, both above and below ground. The solution, he said, was ripe for life if only the right key was used to unlock nutrient-rich accumulations of trace minerals, all of which surely figure in the life process.

Each cell is a little battery. It puts out a current. Deprived of this function because of nutrient shortfall or marked imbalance, the cell dies and deprives living tissue of its service. Murray's remarks were stage setting for those who would read Tom Valentine's book on sea energy agriculture, and they serve us now to preface a body of knowledge that ought not be allowed to die.

The late Ted Whitmer of Glendive, Montana first called my attention to sea solids as a potential plant production fertilizer. He had explored the commercial prospects at considerable expense, and before he passed away in 1996 he counseled *Acres U.S.A.* to add the lessons contained herein to the body of knowledge rescued from oblivion over the decades since the publication was founded. In the final analysis, Maynard Murray was his own best spokesman, the following extract and abstract clearly ratifies.

Listen to this man as he works substance to baffle and teach.

> *Anything living alters its environment for its benefit in order that it may live and reproduce. Life is contained in the cell. Cells vary in size. The largest cell on Earth is an ostrich egg. The smallest is a small bacteria. In the warm-blooded animal, the reproductive cells are the largest and the smallest, the sperm cell being the smallest, the egg being the largest. These cells are able to carry on the process of life.*

Cells need only food from the outside. They can break down complex compounds and synthesize their own body tissues.

A virus, which is smaller than the smallest bacteria, has to live within the cell. Living tissue has to get its food by either concentrating or diluting its environment in order to make its environment a part of its tissue.

All of life is parasitic, exceptions being few. Plants are the exception. Plant life contains chlorophyl or a substitute pigment. Science has identified these pigments by which plant cells can synthesize their own tissue out of simple inorganic elements, the plane of observation defining the locus of organic and inorganic in the life process. Chlorophyl is the pigment in the green algae. The pigment in the retina of the human eye—contained by certain cells—can synthesize proteins, etc., out of angstrom-sized inorganic materials.

Green plants will not use organic materials of the micron size. The plant is not fed organic materials. The soil is the container, and it in turn feeds the plant. It is the supreme function of the soil and its life forms to prepare for plants a suitably sized diet composed of life's diversity minerals, which explains why and how some plants draw nutrients from the air as well as the soil and solutions.

Just as plants require a diet that chemists might style organic—taking the transport mechanism into consideration—animals turn to the mineral box only when desperate, leaving it to microorganisms in the gut to refine and make available nutrient values contained therein.

All this prefaces the introduction of sea solids into agriculture. Sea solids are known as sea salts, and salt is considered more toxic than many mild fungicides used in apple groves. Yet sodium and chloride tied up in many vegetables is perfectly tolerated even by patients on a low salt diet. It has been said that all absolutes are false—including this one. Often this is true with minerals, as is the case with iodine, plant iodine stepping up metabolism, potassium iodine stepping it down.

Nevertheless, Murray does not rely on disclaimers to support his case for sea solids. Pictures presented here tell one part of the story. The rest is supported by experiments and case reports.

Murray is quick to point out that life started in the sea. Human blood is about 25 percent sea water. Fully 85 percent of life on Earth is in the sea. Two elements will not stay in solution: phosphorus and iron. Scientists now consider phosphorus is leaving the land at a tremendous rate. The only return from the sea is via bird droppings, one to three percent going to the sea. In the sea phosphorus hugs the bottom, insoluble, unavailable.

Murray's quest has been to use sea solids, all of them—92 elements included—on acres, orchards, pastures and gardens. Some of the details have become part of the scientific literature. Others are still to be discerned by innovative researchers and farmers. Of special interest is the new hydroponic application discussed here, an application of nutrients that repairs most of the shortfalls of the simplistic hydroponic salt fertilization.

Sea solid fertilization does not excuse farmers from supplying major nutrients, not only the phosphorus (P) discussed above, but the nitrogen (N) and potassium (K) as well. All are best accomplished when the natural nitrogen cycle is working and when the natural carbon cycle is working. Needless to say, the sea fixes nitrogen. The food supply fertilizer is sea water.

Murray reports total success using sea solids on every crop ever tried. His record invites scrutiny, emulation and reiteration. Go to chapter one, then follow the storyline through photographic evidence.

Drama requires backgrounding. This Tom Valentine does by detailing the character of the human health profile and the reason for being of sea solids research in the first place.

"We, indeed, can build up immunity to staph infections, viral and fungal infections in plants," was Murray's parting shot when he came to that Acres U.S.A. conference. "When we grow corn, wheat, oats, etc., and feed them to animals, we see changes." Using animal research with a species bred to get cancer, and feeding them with food grown with sea solids, the first generation cut debilitation from 97 to 55 percent, a significant drop. Through each generation sea-solids food installed resistance to cancer, that is one kind of cancer in mice. Ditto for leukosis in chickens, arthritis in rats, that is, rats bred to get arthritis can be excused from the disease with the foods produced with sea solids.

Farming has to be the beginning of preventive medicine. That is the Maynard Murray, M.D., conclusion, and the start of *Sea Energy Agriculture*.

—*Charles Walters*

Introduction

Animal and vegetable life in the sea is far healthier than similar life on land. Why? Some people believe the buoyancy of a water environment prevents many of the stresses and traumas experienced by creatures living a lifetime constantly overcoming the forces of gravity. Although buoyancy may be a partial factor, it cannot explain why the same species of trout lives two times longer in the saline ocean waters than it does in fresh waters. This curious health phenomenon indicates that the sea provides for its creatures a totally balanced and adequate chemical diet, while fresh waters and rain-washed land masses do not.

Questions posed by the radical health differences between sea life and our landlocked environment have occupied the research efforts of Dr. Maynard Murray, a practicing physician and physiologist, for 45 years. This book is the result of his lifelong search. His work has opened doors to a provocative new arena of science and technology called "Sea Energy Agriculture," and it is quite possible that this new field of learning will lead to the end of disease and famine.

Such a prospect is most encouraging since our world sits on the verge of a terrible crisis in agriculture and food production. Of course there is still much more to learn, but Dr.

Murray's efforts have established a firm foundation for future research. When preparing to announce his findings to the world through the publication of this book, Dr. Murray said:

> *Life is far too short for one person to selfishly guard any new facts he discovers. Therefore, I am revealing all I have learned even though some of the data is not yet complete. Many minds are better than one, and it is my ardent hope that from this beginning more enthusiasm will be generated which will bring active, probing minds into the field. The results of my beginning research must be amplified and technologically developed in order to best serve mankind.*

A huge portion of our aggregate lifetimes and total resources is spent combating illness and trying to withstand the ravages of age. It is paradoxical that despite the great variety of foods we have developed to nourish our bodies, we still suffer degenerative disease and fall prey to the aging processes long before the optimum life-span for humans. It has been said time and again that we are what we eat. This truism complements the simple fact that although we Americans have greater abundance, and perhaps a more balanced diet than most of our primitive forebears, our intake of vital, life sustaining elements is woefully inadequate. The people of the United States are the best fed, chemically starved people in the world.

Statistics are difficult to keep accurately in a nation as large as ours, but in recent years statistical studies of disease have improved considerably—and the data revealed is frightening. There is a tremendous increase in the frequency of chronic and metabolic ailments. Dr. Murray's research clearly indicates that the reason Americans generally lack a complete physiological chemistry is because the balanced, essential elements of the soil have eroded to the sea; consequently crops are nutritionally poor, and the animals eating the plants are therefore nutritionally poor. Our scientific efforts to isolate and synthesize what we learn to be the essential properties of fertilizers are impressive, but man's methods apparently have not satisfactorily duplicated nature's methods. Something is obviously missing.

Because he was a scientist, Dr. Murray had a great respect for what our technology has accomplished, but he warned that we must accept a junior partnership with nature. "If we do," he stated, "she may allow us to survive. If we do not, she undoubtedly will eliminate us just as surely as she has the brontosaurus, the wooly mammoth and all the other creatures who once also 'ruled the earth.' To join in this junior partnership we must alter the way we are growing our food, the way we are protecting our plants from pests and disease and the way we process our foods."

Many prevailing beliefs about soil and plant growth are erroneous and must be changed. Dr. Murray's experiments proved that an adequate supply of food can be developed if man recycles the sea. Beginning in 1936, Dr. Murray experimented to determine what elements in the sea harbor the secret of healthy plant life, which in turn contributes to the health of the animals who consume those plants. He became interested in hydroponics, the art of growing crops in a liquid solution without soil, as a means of controlling which elements would be present in the nutriment available to the plants in his experiments. He tried solutions made from evaporated sea salts to determine what means of balance was available in the natural sea water and what effect it would have on plants. Sodium chloride, the major component of sea salts is normally toxic to plants. However, Dr. Murray discovered a method to prevent the salinity from affecting the root structure of the plants he was testing.

From the start, Dr. Murray's sea salt experiments produced excellent results, and it has now been conclusively proven that the proportions of the trace minerals and elements present in sea water is at optimum for the growth and health of both land and sea plants. In 1954, controlled crop experiments were conducted. Corn, oats and soybeans, three staple feeds, were used. Ten acres of sea salt and control corn, ten acres of sea salt and control oats and six acres of sea salt and control soybeans were grown. Subsequently the produce was fed to animals under controlled conditions—four parts corn, two parts oats and one part soybeans. Not only were the crops superior to the control crops, but the effects on the physiology and pathology of the animals fed the sea salt produce were delightfully amazing. For example, chickens, pigs and cattle fed sea salt produce reached maturity much sooner than control animals, and all resisted diseases common to their species better than control animals. Experimentally fed pigs carried the benefits into a second generation and there were no runts in the litters, which is something that "always happens" in a litter of pigs and is a sign of malnutrition.

Throughout this book Dr. Murray's research is brought to you in detail. Who is this man whose theories and research portend a near revolution for agriculture? Dr. Murray received his bachelor of science degree in 1934 and graduated from the University of Cincinnati Medical School in 1936. Two additional years of post graduate study in internal medicine followed, then an additional three and one-half years for his specialty, ear, nose and throat.

With a voracious appetite for learning, Dr. Murray taught physiology and directed a number of experiments at the University of Cincinnati between 1937 and 1947. Meanwhile he attended night school to study law and trained in medical hypnosis under Dr. Mullenhoff. He was a member of the Association of Medical Hypnotism, the New York Academy of Sciences, the American Association for the Advancement of Science, the American Medical Association, Chicago Medical Society and the Illinois State Medical Society.

In 1947 he moved to Chicago and established a private practice. For the following 25 years, Dr. Murray practiced his specialty, carried on extensive experimentation in sea solid fertilization, and authored at least 20 articles, which appeared in national and international medical journals.

He later lived in Ft. Myers, Florida, where he was Medical Director at Sunland Center and was voted boss of the year, 1975-76, by the Medical Assistants Association. He was engaged in commercial hydroponic tomato farming using his patented sea solid technology.

While presenting his ideas, Dr. Murray often criticized current practice and theory. He did not relish being critical, and he certainly did not convey any gloating self-righteousness. The criticism in this book is meant constructively, and if the result of his lifelong work eventually leads to less suffering and illness for mankind, all argumentativeness will be worthwhile.

—Tom Valentine

Chapter 1
Longevity Versus Health

Without health, life is not life; it is only a state of languor and suffering—an image of death.

—Rabelais

The health and physical conditioning of civilized Americans is, in the main, alarmingly poor. Moreover, it worsens each year. I make this assertion as a physician who has practiced for more than 40 years, and I make it despite the media propaganda that we are living longer and better than ever. You and I, fellow Americans, hold the dubious distinction of being among the sickest of populations in modern society.

My intention is neither to stir dissension among the ranks of the healing professions, nor to sabotage the efforts of modern medicine, rehabilitation, pharmacy and allied fields of care. My true intent is improvement.

This "heresy" of mine is shared by health food advocates, organic gardeners and others who are considered to be within the ranks of the "quackery" fringe. It is an unfortunate result of social conditioning when advocates of a better way are suppressed by a powerful, economic dogma.

Although I am critical, I do not belittle my own profession which has done so much to prevent such terrible diseases as diphtheria, smallpox, tetanus, cholera, leprosy and typhoid fever. I am critical that our so-called progress has been primarily reactive in forestalling death, not proactive in alleviating causes, especially in the areas of chronic and metabolic disease.

More than 100 million cases of chronic or long-running illness and disablement afflict U.S. citizens today. That's nearly half the population and many of these cases afflict the very young. These statistics are even more alarming if we take into account less disabling diseases such as dermatosis, chronic migraine headache and dental disease. And finally, the topping on the unsavory morsel of medical fact is the distressing truth about infection. Despite our wonder drugs, steroids, sanitation standards and general medical wizardry, the United States has one of the highest infection rates per capita of any society, perhaps even higher than India, a nation that we are trying to help with our medical know-how, sanitation practices and pharmaceutical advances.

America the beautiful! A magnificent nation ranked first in the world in many categories that characterize advanced civilization. We take pride in our public education and our scientific/technical achievements and we advertise in glowing terms the tremendous advancements we have made in medical science. I agree, America is the best—but it can be so much better. We've got to stop kidding ourselves with distorted health statistics and do something about our shockingly poor national health record.

There is no excuse for 97 percent of Americans having some kind of chronic dental disease. That's a terrible statistic, it's damn near everybody. Visual impairments requiring corrective lenses belong to nearly half the population—especially when one considers the great number of people who need glasses but have not yet become a statistic. Every year more than a million people die of heart and vascular diseases, and one out of every 16 Americans aches with arthritis and rheumatism. And think seriously about this one; there are more than 250,000 children suffering from rheumatoid arthritis.

Are we kidding ourselves with longevity statistics and "war against disease" statistics? You bet we are! That last item, juvenile arthritis, for example, draws the neat statistic from the experts that we are doing so well against this disease that only one in ten of the victims is permanently crippled. The others merely suffer pain and partial crippling.

Our medical statistics equate the increase in the average life span with good health. This kind of reasoning makes appealing news copy, but it reminds me of the statistician who drowned in a river with an average depth of only one foot. Such propaganda is deathly deceiving. This specious reasoning can lull us into a feeling of unwarranted well-being.

I promise not to belabor this negative approach much longer, but it seems we Americans, collectively, are a lot like the Mexican burro that requires a smack between the eyes with a two-by-four plank before attention is focused. What our health statistics are saying is this: because insulin increases diabetic life spans for 20 years, those diabetics are in "good health". Because a cancer patient was treated with radiation, chemotherapy and surgery, he enjoyed "good health" during the extra two or three years of prolonged life such medical wizardry afforded him. Consider a patient who is paralyzed by a stroke, losing the power of speech, but kept alive in a wheelchair for three years, thereby statistically lasting longer. And how about the cerebral palsied child who would have died in infancy but was "saved" by the wonders of medicine only to die after spending 20 years in bed in constant need of nursing?

One needn't be a qualified medical doctor to see what is happening. Our great medical skill concentrates on prolonging life despite the afflictions. It is erroneous, therefore, to conclude that statistics increasing life expectancy are tantamount to good health.

Whenever life expectancy is considered, it is more logical to assume a biological viewpoint rather than an insurance company viewpoint. It has been learned that throughout nature the full life expectancy of an animal is based on a multiplier utilizing a growth period to maturity concept for the particular species. The most conservative calculations for this growth period multiplier is six while some researchers insist it is ten. If an animal is fully grown at one year, its full life expectancy is six times that, or six years. That, as you can see, is a conservative estimate. Since man does not achieve full maturity until he is about 20 years of age, his full biological life expectancy would be 120 years. How many Americans attain even two-thirds of 120 years?

That this estimate of a life expectancy exceeding 100 years is realistic *is* attested by various small tribal groups around the world exhibiting centenarians who are vital, active and not at all uncommon.

All right, we've stated a problem. Wherein lies a solution?

I suggest we strive to get back aboard mother nature's merry-go-round. Unlike the vocal arrogance of mankind, smug boasts are not heard among the sounds of nature, and there is no complex, scientific jargon among her simple truths. Nature is immense and minute, barren and abundant, silent yet teeming. Amidst all this magnificence and power nature is humble. She speaks most often in straightforward, single syllable terms—curt, chilling, impersonal and final. Sometimes she acts with a seeming cruelty that appears appallingly whimsical to man. Yet, in spite of her callous behavior, nature is kind and generous. She feeds, comforts and nurtures all her living creatures with amazing efficiency and consummate skill.

Because we humans have qualities of mind that set us apart from the rest of nature's creations, we tend to grow arrogant and believe we are above nature. Even when nature coldly reminds us that we fit into her biological scheme of things, we humans have the damnable ability to blot the essence of these reminders from our thinking.

Life on this water-doused, air-covered planet exists only because there is death. With the exception of those species of plants blessed with chlorophyl, all living creatures have sustenance only because some other living creature before them died. Some people see it as a vicious cycle, others see it as a perfectly balanced plan with the survival of the fittest, the most worthy. I see nature as a magnificent struggle—all that is alive struggles against the adversities posed by wind, sun, water, cold, hunger, thirst, light, time, gravity and space. From microorganisms, which are plant life, to elephants and whales, there is a continuous battle to gain control of earth's elements. In every cubic inch of earth's outer crust, in its oceans and, to a lesser degree in the air above, the relentless fight for control of the elements never wavers.

Although the vegetable and animal kingdoms vie for the essential elements, they each owe their very existence to one another. The first order of life is breath. Animals breathe the oxygen the plants produce and the plants utilize the carbon dioxide the animals exhale. It's an efficient system that only man in his arrogance thinks he can shortcut or circumvent.

It isn't my purpose to try outlining a course in elementary biology, but these essentials were either lost or ignored as we probed more deeply into microbiology. All animal food, including that which rests in our refrigerators at this very moment, originates in the cells and tissues of green plants. From the one celled algae to the giant sequoia tree, green plants alone possess the power of photosynthesis, that remarkable ability to convert lifeless, inert elements of the earth into living, nourishing foodstuff. Using light-energy, the plants hook up a carbon atom to the inorganic elements and make them organic. Animals eat the plants, other animals eat the plant eaters and, as bigger other animals eat the smaller, the procedure goes on up to man who claims dominion over the other kingdoms of nature.

However, at the end of the upward thrust of nature's chain, the tiniest of organisms, single celled non-green plants called bacteria, fungi or viruses, hold sway over man. Decaying organic cells go back to earth for recycling. Man is leveled, usually before his biological time is up, by an as yet unexplained aging process which is managed by the tiny creatures.

Since we humans are composed of cells which vary in complexity, kind and number, we can live only insofar as we nourish those cells. In short, we are what we eat. The cell function in the animal body is highly specialized nourishment which enables it to perform its function. Hence, the food the cell receives supplies not only the requirements for building and maintaining physical properties, but also this nourishment must supply that specific substance necessary for the cell to perform its particular function.

Without knowing how nature does it, we do know for sure that specific requirements for cell nutrition exist and must be adhered to for optimum health. For example, we know that calcium is important to the parathyroid cells; iodine is inimitably associated with the functioning of the thyroid cells; iron is required by the red blood cells; sulfur is necessary for some of the functions of the pancreatic cells; copper is essential to the liver and zinc appears to be associated with the function of the gastric mucosa (digestion) while bromine is required by the pituitary cells.

There must be hundreds of health and nutrition books that stress how any absence or deficiency in these specific food essentials results in an inadequate cell-group function. One cell group lagging behind can have an adverse effect on another cell group and the next thing you know, the entire conglomeration of cells operates in a weakened state and becomes fair game for hungry hosts of parasitic microorganisms that are always hoping for an easy meal. Pounced upon weakened cell groups are usually referred to as disease.

Specialized feeding is not only true among groups of cells. It is also observed in all realms of life. It is apparently to effect a balancing of life struggles that nature adheres to this principle. For example, the koala bear of Australia depends upon certain types of eucalyptus leaves for nourishment; the gorilla of the African highlands depends upon the tender hearts of wild celery; the giraffe reaches high for acacia leaves. There is even a variety of deer which lacks a gall bladder and therefore must eat poisonous coral plants to live. Still another example is the elk's need for cattails or plants carrying traces of gold in their structure. And the insect world is a potpourri of specialized eating habits and parasitism.

There is a point in this rehash of generally known information, and it is made when we earnestly consider the contenders for the title of dominant life form. Mankind, insects and certain small rodents are doing very well on this planet, but by far the most dominant life forms are the bacteria, fungi and viruses which are microbes. These adaptive creatures outstrip all others in their ability to thrive and propagate in the extremes of climate and environmental change and food availability. Microbes permeate every living thing in every conceivable location on earth. Multiplying by simple cell division that can be amazingly complex in some species, their existence is threatened only by extremely high temperatures, certain chemical compounds and each other. Even a lack of nourishment is no great threat to many microbes because they may hibernate for long periods of time. Some microbes have "returned to life" after being uncovered by archeologists in sites said to be buried thousands of years ago.

The importance of microbes cannot be ignored by man, for not only do they feed upon man, they are the apparent primary cause of his aging and death. Microbes epitomize the principle of selective feeding and they have some of the most peculiar and fascinating food preferences and eating habits.

We have learned to utilize some of the unique feeding habits of microbes in commercial enterprises such as in fermentation. The microbes performing these specific tasks do so because they have developed specialized tastes and metabolic reactions to the materials they ingest. Their eating habits sometimes transform a particular material into a different structure often giving it bouquet, hardness or softness. The affinity some microbes display for specific body parts within humans is truly amazing. One must conclude that a certain section of body tissue produces or harbors the particular food element a particular microbe deems necessary for existence. The microbes search out this food from many available sources with apparent relish. For example, the bacterium which causes diphtheria produces a toxin which attacks the myocardium or middle layer of heart tissue; actinomycosis is caused by a fungus showing a strong preference for the lymph glands of the neck. Streptomycin is an antibiotic derived from a fungus which attacks the tiny balance mechanism of the ear. Yet it ignores the cochlea, which is the hearing cell(s) within the same structure. On the other hand, dihydrostreptomycin attacks the cochlea and leaves the associated tissue in the ear alone. The virus strain causing rabies has a special affinity for nerve tissue in the central nervous system. Epidermophytosis is caused by a fungus attacking skin tissue, and curiously enough it prefers the skin of the thighs and feet. Subacute bacterial endocarditis is the medical term for a disease affecting the inner layer of heart tissue caused by the eating habit of the *Streptococcus viridans*, a bacterium. And the list goes on and on—one of the reasons medical science is so fascinating.

Selective eating habits of microbes is part and parcel of nature's progressive and balanced feeding system. It is fundamental biological law. It is a principle that *must be appreciated* if we are to seriously tackle the problems of aging and disease. Here I want to digress for a moment to make another unpopular medical assertion. I am convinced that all disease stems from a weakening of the organism and subsequent parasitic infection. There is a growing school of thought that attributes the *cause* of disease to psychological upsets or emotional stress. There is abundant thinking that mental illness is a non-physical disease. I am convinced that these psychological tenets are now being accepted simply because no apparent biological cause can be found. In truth, no human who is consulting a physician for treatment is free of parasitic microbic life. Emotional stresses may complicate matters and may even appear to be the cause of the malady, but first and foremost that body has been inhabited by a host of microbes producing toxins and degenerating cells. The problem is to understand infection, its exhibitions, shape, forms, manner, presentation, reactions and insidiousness. When this is fully understood by the diagnostician the tendency to resort to ambiguous, psychological diagnoses will be greatly reduced. All organs, the brain included, are affected by microbic life. The key to controlling disease in my studied opinion, is a healthy and optimum functioning conglomerate of cell groups which is possible only through optimum nutrition.

My digression continues because it is so very important that we all understand some of the things that are happening in the professional world of medicine. Recent outbreaks of staphylococcal epidemics in hospitals throughout our country and the various waves of influenza viruses sweeping across our land stress the constant menace of infection by microbes. These illustrate the massive power which microbes possess. Though this is a power that must not be underrated, a laxness has developed in the minds of many physicians and laymen. The medical profession has created a sense of security based upon an illusion that wonder drugs and antibiotics can solve everything. Unfortunately, despite drug industry propaganda, these new medicines are fraught with shortcomings, and the long-term effects may prove them more harmful than beneficial. It may be said that I'm just an old country doctor. Fair enough, but I share the view of Dr. Haskell Winestein and Dr. Maxwell Finland who are eminently more than mere country doctors.

Dr. Finland and his colleagues examined Boston Hospital records covering a period of 24 years in order to evaluate the long-term results of wonder drug therapy. They learned that wonder drugs had reduced the death rate from infection caused by pneumococci and streptococci, but there had been an increase in deaths due to infection from bacteria which were previously considered harmless. Reliance on antibiotics to combat infectious disease is to "live in a fool's paradise," noted Dr. Finland.

My years of experience, both as a practicing physician and a researcher, have convinced me that health versus disease is based upon the story of eating and killing—a tale of growth and cell building by the ingestion of elements taken from other creatures. This must be so because it is the story of life and the story of disease, decay and death.

Shakespeare broached the subject a half century prior to the invention of the microscope and fully 200 years before Pasteur when he wrote:

> *Your worm is your only emperor for your diet:*
> *We fat all creatures else to fat us.*
> *And we fat ourselves for maggots.*

So goes the eat and be eaten world in which we live. And if we wish to live well and long, we must concentrate with the utmost seriousness upon proper nutrition. Of necessity, this attention starts with nutrients balanced in the sea.

Chapter 2
Soil, The Source of Health

Go to the ruins of an ancient and rich civilization in Asia Minor, Northern Africa or elsewhere. Look at the unpeopled valleys, at the dead and buried cities, and you can decipher there the promise and the prophecy that the law of soil exhaustion holds in store for all of us.
—V.G. Simkovitch

Has it occurred to anyone that rain—the daily phenomenon we regard as essential to life—may, in our present stage of agricultural evolution, be responsible for some of the diseases which afflict humanity? Rain not only gives, but takes; not only blesses, but condemns.

Chemically speaking, rain is simply distilled water. It has pelted the earth since the beginning of time, helping to chisel the ground's features, which we recognize as landmarks. Like all liquids, rain seeks its lowest level after it has reached earth. If enough accumulates in one place it forms streams and rivers in the quest for the level of the ocean. When the cutting and carrying capacities increase by accumulation, rain then removes the top soil by erosion, carrying it to lower ground and eventually to the sea.

Much has been said and done about soil erosion and the loss of topsoil since these conditions can be readily observed and easily understood. But what of the vital constituents which make up the topsoil? How much of these still remain in the topsoil which has not been carried away? Have we "conserved" and protected the soil from erosion? Is there any relationship between possible "unseen" soil losses and the afflictions of humanity?

Rain plays an integral role in the growth and development of all living things by contributing water, nature's universal solvent and chief building material for all living matter. Because of its capacity for dissolving elements, rain makes substances, which otherwise would remain inert in the earth's crust, available to all living creatures. Since plants cannot absorb elements from the ground unless they are in solution, the action of rainwater dissolving minerals and transporting them from place to place is vital and necessary for the perpetuation of life. Plants and animals owe their existence to it and without it all earthly forms of life disappear. Thus, rain gives.

Rain, however, also washes the soil and cleanses it of many elements, carrying them off to the sea. Whenever it rains, the soil is leached of some of its vital constituents so necessary for plant nourishment and life. Thus, rain takes.

That the process of soil leaching occurs is not a cause for alarm. What is alarming is that soil leaching in America is increasing and accelerating at a tremendous rate leaving our greatest natural resource, the soil, badly depleted.

When man discovered he could control and even accelerate plant growth, he began to till the soil although the first attempts were sporadic, individual and local. As time passed, man organized his agricultural endeavors and settled on the lowlands along banks of rivers and streams. Here a supply of water was available and the soil, rich in river gleanings from upper regions, had been deposited. To meet the food demands of growing settlements and eventually great cities that flourished on these river banks, the soil was worked to a greater and greater extent Cultivation boundaries grew beyond the environs of river valleys and, in the process of expansion, man invaded and removed the protective water-holding cover afforded by naturally occurring plant life. The jungle was cleared, forests denuded and the plains were plowed. Had the processes of natural plant life on earth been left entirely to nature, rain would not have become an important source of soil depletion. Instead land was cleared for houses to be built, roads to be laid and crops to be planted and the great agricultural boom set in.

Within the last century, a large percentage of the arable land in the United States has been stripped of its original protective plant covering. To the greatest extent, the land has been converted to produce food at an overwhelming rate for an ever expanding population and the point has been reached where we force the land to produce larger quantities every year. Following are some comparative statistics for crop yields published by the United States Department of Agriculture:

Corn:
    1900-1909   average yield was 27.3 bushels per acre
    1950-1960   average yield was 43 bushels per acre
                          Increase in production: +58%

Wheat:
    1919-1929   average yield was 12.7 bushels per acre
    1950-1960   average yield was 19.7 bushels per acre
                          Increase in production: +55%

Rice:
    1919-1929   average yield was 1,818 pounds per acre
    1950-1959   average yield was 2,797 pounds per acre
                          Increase in production: +54%

Oats:
>1900-1910   average yield was 28.9 bushels per acre
>1950-1959   average yield was 36.2 bushels per acre
>> Increase in production: +25%

Sugar Beets:
>1913-1923   average yield was 9.9 tons per acre.
>1950-1959   average yield was 16.4 tons per acre.
>> Increase in production: +66%

Vegetables, fruits and other field crops also show similar increases in yield per acre and stand as a tribute to the productive ingenuity of soil scientists and farmers. Nevertheless, as a result of this intensive and extensive exploitation of our land, we have depleted the soil in a three-fold manner:

1. *Physically*, by persistent and continuous plowing and cultivation of the land, thus facilitating the leaching of vital soil nutrients through the action of rainwater.
2. *Biochemically*, by the continuous planting of food crops that draw mineral substances from the soil to build plant structures.
3. *Economically*, by failing to replenish the soil with all elements taken from it in the agricultural process, instead reinstating only the three to six primary elements.

Consider the soil as a gigantic mine filled with precious stones, rare metals and elements of all sorts, teeming with macroscopic and microscopic life. Though it varies from place to place, if one were to analyze the chemical composition of the well-formed soil that had not been depleted by leaching and crop production, one would find relatively large amounts of nitrogen, phosphorus, sulfur, potassium, calcium and magnesium. These are the primary elements considered necessary for plant development and growth. To a lesser degree and in varying amounts, soil would also contain copper, zinc, boron, molybdenum, vanadium, manganese, cobalt, iodine, chlorine, iron, fluorine, sodium, barium and other elements. Because relatively small amounts or traces are found in the soil, these secondary elements are called "trace elements."

Since the primary elements as listed above are considered the mainstays of the plant's nourishment and the building blocks of the plant's main structures, the trace elements must be considered vital to the subtle good health of the function of the plant. This contention can be illustrated by citing specific examples of plant diseases found to be the result of trace element deficiencies in the soil.

As early as 1928 "grey speck of oats" was found to be caused by a deficiency of manganese. Since then, similar diseases such as chlorosis in tomatoes and Pahala Blight, a disease in sugar cane, were found by scientists to be actually caused by a deficiency of manganese which allowed infection in the plants by fungus. A virus disease of camellias which produces yellowish splotching on the leaves and white mottling of colored flowers has been found to be caused by a deficiency of iron in the soil.

Rosette is a disease that affects many varieties of fruit trees and is due to a zinc deficiency. Exanthema, another disease manifested in fruit trees is caused by a copper deficiency. Rutgers University scientists have found that molybdenum deficiencies can cause plant diseases while boron, an element which has also gained recent prominence as a fuel additive, is known to be essential for the general health of the cauliflower and other plants.

These cases are just a few of the examples of plant diseases which have been attributed to the lack of specific trace elements and plant and soil scientists frequently discover others which appear to stem from a specific deficiency in one or a combination of elements in the soil. The loss of these elements, you will recall, is in part the result of rainwater's leaching action and is greatly increased in effect by our practices of intensive soil cultivation.

If a soil is like a mine with its myriad of elements, then whenever man tills it, plants crops and in other ways engages in farming, he is for all practical purposes engaged in the business of mining. Like a miner, the farmer breaks the earth with digging tools, but instead of dynamite, he plants seeds to loosen the minerals and elements from their holding matrix. And finally, he carts away his minerals in the form of food rather than ore.

Whenever a field of corn is planted, the soil must be mined of 48 elements since a healthy ear of corn contains that many elements. Wheat and oats both utilize over 36 elements apiece while soybeans, apples, pears, peaches and other fruits each take out over 30 elements from the soil. Most vegetables, including peas, potatoes, tomatoes, carrots and lettuce take more than 25 elements each from the soil. These considerations, while impressive enough in themselves, would warrant only passing comment were it not for the fact that our soils are showing signs of exhaustion. The frightening thing about this "mining" enterprise is that the farmer will replace only three to six of the total number of elements removed from the soil by crops. It is obvious that this plunder cannot continue indefinitely without serious consequences, most notably in decreased plant and animal nutrition.

The soil in the United States, overworked and underfed as it is, complains of the burdens being placed upon it and manifests its discontent by producing improperly. Diseased plants and sick animals are the result of failure to replace all the elements taken out during the growing cycles. From time to time examples abound in California where, even when climatic conditions are favorable, no fruit appears; on the pineapple plantations of Hawaii where fruit is decreasing in size; in the peach orchards of Michigan where pests and disease are common and increasing; in the rice paddies of Arkansas, Mississippi, California and Texas where disease is perplexing; and in the pea patches of the middle west where pea sickness is not uncommon. Examples exist throughout the United States and wherever intensive farming is practiced.

With science to show him the way, the farmer learned how to inveigle the soil into greater and greater productive efforts on less and less vital substance. Science first developed a breed of crop which produced an abundance of seed or fruit. What mattered most was the quantity because farm remuneration is paid primarily on the gross production basis. This practice reserves a secondary place of importance for crop (and seed) quality. Consequently, 80 bushels are always better than 60 bushels even though 50 percent of the former is air while only 10 percent of the latter is. The national farm motto has become "The Number of Bushels Produced."

Hand in hand with the development of high production and uniform sized seed was the contrivance of mechanical devices which greatly increased the planting, cultivating and harvesting capacities of the farmer. These seeds, when sown by machine, must be uniform in size and shape and the fruit, in order to be reaped by machine, must also be uniform in size and shape as well as ripen simultaneously to be harvested economically. Consequently, most major food producers utilize only a few varieties of seed per crop. Is it any wonder then that our crops are so much more vulnerable to destruction by virus, fungus or other disease? This vulnerability stems from the very narrow base of genetic uniformity which aids the efficiency of agriculture but opens the crop up to widespread pathogenic disease. For example, the corn blight which occurred during 1970 diminished the United States crop by approximately 700 million bushels and included over 90 percent of all hybrid corn grown in the country that year.

But probably the most important contribution to high farm yields and continuous plant production was the practice of replacing the major elements for a plant's nutritional needs in the soil and fortifying them with heavy doses of nitrates. Thus, soil production was maintained and even increased in spite of starving the soil of vital trace elements necessary for the sustenance of healthy life.

Accordingly, the growing of food of increasing bulk, filled with fiber and air and devoid of vital substance cannot help but diminish the strength and vital substance of those who ingest it.

Nature, while temporarily outsmarted by the farmer into growing plants on an incomplete soil, has not been outdone. In return, from nature we have received the following:

1. A glut of food that is great in bulk but low in vitality.
2. A lack of storage space for this glut.
3. Increased incidence of disease in our crops.
4. A soil badly depleted of trace elements.
5. Animals lacking vital elements essential to their health.
6. Rivers rich with elements fed them by the draining rain, which they carry out to sea.

This is the promise that the Law of Soil Exhaustion holds in store.

Chapter 3
Trace Elements
in Nature's Balance

It is the close observation of little things which is the secret of success in business, in art, in science and in every pursuit of life.

—Samuel Smiles

Approximately 70 percent of the Earth's surface is covered by oceans and other collections of water. Protruding from these vast waters are pinnacles of land called continents. Over an estimated two billion years, the land has been worn down by rainfall, which washes the various soil elements out to sea. Thus, it is apparent that the sea is an enormous receptacle of the former chemical richness and balance that once supported life on land. Although it is not possible to know the exact rate of chemical denudation of the land masses over geological time or even at present, from recent estimates it has been determined that the rate ranges from six tons per square mile for Australia to about 120 tons per square mile for Europe. On a worldwide basis then, about 4 billion tons of dissolved material are carried to the sea by rivers each year. The most soluble elements are first picked up by rainwater and that is the reason why sodium chloride (common table salt) is so scarce on land, yet abundant in the sea. In several million more years, nature will succeed in completely eroding the land masses so that the sea will once more cover the earth and the cycle will be complete. Concurrently, geological forces will again raise land masses which will exhibit the rich chemical balance of the present sea water. This new state of balance will be temporary when taken over an extended time span since the same cycle of erosion would begin again.

Sea water is the most ancient natural solution on earth and, in my opinion, it is the most ideal physiologically. The disease resistance of plants and animals in the sea is remarkably different from disease resistance in land animals and comparisons between animals of the same or similar species are most interesting. For example, fresh-water trout all develop terminal cancer of the liver at the average age of five and one-half years; cancer has never been found in sea trout. It is also known that all land animals develop arteriosclerosis, yet sea animals have never been diagnosed as arteriosclerotic. Investigators have also established the startling absence of disease in the sea, citing not only the absence of "chronic" disease forms, but especially the general vigorous health of sea animals that has apparently lengthened life many times in comparison to similar land species. These longevity differences are especially evident in such sea mammals as whales, seals and porpoises who have identical physiological systems with the majority of land animals important to man. And the major differences between sea and land life appear to be attributable to the superior food chain of the sea!

The top soil of land is characterized by elements in a colloidal state, defined as "a gelatinous substance which when dissolved in a liquid will not diffuse readily through animal or vegetable membranes." The sea is characterized by elements in a liquid crystalloid state, defined as "a crystallizable substance which, when dissolved in a liquid will diffuse readily through vegetable or animal membranes." Unlike the colloid state of top soil on land, the liquid crystalloid of the sea retains only the amount of each element that maintains a consistent chemical balance. Hence, excessive amounts of any given element(s) will drop to the bottom of the ocean where it can be taken up only if the plant and animal life have depleted that element from the sea water solution. Thus the chemical balance is maintained.

The colloidal state of the land causes the opposite effect. When an element is leached from land, the resulting imbalance causes either a blocking of the other elements present so they cannot be taken up by the plants, or a substitution of some other element (for the one leached) takes place. As more and more of the top soil elements were leached away, man began to put back manure, decayed foliate and dead animals for fertilizer. In the process he had returned the elements to soil in the same proportion as they had been cropped out. In modern times, agriculture has begun the process of adding the basic elements of nitrogen, phosphorus and potassium plus lime (calcium chloride) in large amounts which initially has caused the yield from crops to increase. As has already been pointed out, however, there is a growing evidence that excessive buildup of these four elements blocks the uptake of vital trace elements. In essence this means that leaching away of elements and excessive application of the four macro elements to crop reduced soil have seriously weakened our physiological food-nutrition supply to the point where it is amazing that we are able to function at all. It is no wonder that disease constantly attacks the various land organisms, including humans, in an attempt to naturally recycle the elements so that a fresh start can be made.

In the sea, by the very nature of its liquid crystalloid state, there is no occurrence of blocking or need to substitute elements. All elements of the atomic table are in solution of consistency, balance and proportion, available to all sea life. The sea plants that ingest inorganic elements and thereby begin the food chain always have the same chemical solution to feed on so that their chemical analysis is always identical from one sample to the next. The extreme opposite is true on land where even plants that are grown a few feet apart exhibit chemical differences, especially evident in the micro or trace elements. Given the consistent chemistry of the sea plants, there is never a need to attempt developing "disease resistant strains" as in land seed hybrids because sea plants are always disease resistant.

Further evidence of this consistent chemistry is found throughout the food chain in the sea when we note that animals feeding on a sea plant diet are also consistently balanced. These facts are rendered conclusive when comparisons are drawn between sea and land life. In land animals, for example, the range of iodine numbering in fat is tremendous and packing houses have found differences in animals from the same as well as different farms. In addition, while chemical analysis of muscle tissue of the whale and porpoise is always the same, the same analysis on land animals varies considerably from animal to animal.

An article appearing in *Science News* (Vol. 100, August 14, 1971) entitled "Trace Elements: No Longer Good Versus Bad," indicated the dramatic changes in interest in the topic of trace elements and health by the scientific community over the last ten years. This article points out that only a dozen or so trace element laboratories existed in the United States by 1966. Dr. James Smith, Chief of the Veterans Administration Hospital Trace Element Research Division in Washington, D.C. now estimates that there are over 50 laboratories in the U.S. devoted to working on trace elements and their role in physiology. Research is also being conducted in various European countries, the Soviet Union, Egypt, Iran and Australia.

One of the breakthroughs of major proportions has been in the awareness that a particular element can be essential to physiology at a minimal level although it can be toxic at a higher level. Until just recently all of the research emphasis was placed on determining if an element was toxic and on symptoms of toxicity rather than on considering the quantitative amount and chemical state of the element when it was ingested. As is well known, sodium chloride is used universally as table salt in the inorganic form. Equally as well known in the scientific community is the fact that an excessive amount, such as four or five teaspoons of table salt, ingested at one time is potentially lethal to human life. The use of salt was a recognized method of committing suicide practiced by the Chinese in ancient times. Additionally, it can be shown that an excessive amount of any element is toxic and even a small amount, if ingested by humans in inorganic form, may very well be toxic. As earlier described, people can utilize inorganic salts or elements only by having plant life in their intestines in the form of bacteria to hook up the inorganic element with a carbon atom so it can be transformed into an organic form. It is also interesting to note that many pregnant women and often people with heart disease, etc., are restricted to a salt free diet by physicians. Although one stalk of celery has as much sodium chloride in it as one would normally use at a given meal time through the salt shaker, salt free diets do not exclude celery. The obvious reason is that sodium chloride, per se, is not toxic; it is only sodium chloride in the inorganic state that produces toxic effects.

Crops and Soil Magazine (Vol. 13, No. 7, April-May, 1961) carried an article entitled "Animal Health" by W.H. Allway, ARS, USDA, Ithaca, New York, from which the following quotation was taken: "Thus it may be more effective and efficient to supply certain trace elements to the livestock through the fertilizer-soil-plant route, rather than add these nutrients directly to the animal feed." This statement was based upon the observation that "increasing evidence indicates that various chemical compounds in which the trace elements may occur vary in their effect on animals."

Although only twenty elements (or minerals) are known to have a specific role in human physiology, several more are known to have beneficial effects in the physiology of plants and animals. The heavy metals, i.e., lead, silver, gold, cadmium, mercury, antimony and aluminum among them, have a suspected positive role and even known poisonous elements such as arsenic can be beneficial in some animals if they are ingested in organic form and in the trace amount. Finally, the *Journal of the American Medical Association* (Vol. 201, No. 6, August 7, 1967) has reported that William H. Strain, Ph.D. and Walter J. Pories, M.D. of the University of Rochester School of Medicine and Dentistry are investigators who champion the position that no element presently can be ruled absolutely unessential to humans. In short, specialists in trace elements generally agree that more trace elements await discovery as dietary essentials in various animal species and possibly in man.

Since it is true that major work is being done on the physiological role of trace elements and no element has been ruled out as possibly being important in physiology, why then did I become interested in the use of whole sea water as a fertilizer? The answer lies, at least partially, in the fact that while some 20 elements have been determined as having a role in physiology, there remain the additional 72 elements which make up the atomic table.

It has been estimated that a definite physiological role for a particular element is newly discovered on an average of one every 10 years. Thus, it is apparent that we may have to wait for five or even six hundred years before all are discovered unless the rate of discovery is markedly increased. The very nature of the scientific method precludes that the researcher is not a generalist so the process only allows isolation of one variable at a time in order to identify that variable's specific role. I am not seeking to disparage the work of men in such fields as soil science, plant physiology, animal husbandry and medicine in general, however I am suggesting that we simply cannot wait for the inferred number of years for every remaining element to be identified and its role in physiology to be specifically defined! For example, since only a few of the enzymes have had their necessary trace elements identified, only around nine trace elements are listed under "Recommended Dietary Allowances." However, since thousands of enzymes have been identified, there are undoubtedly thousands more enzyme-trace element joint functions remaining that must be isolated and described. The article "Trace Elements: No Longer Good Versus Bad" describes such action as follows:

> *A trace may stick to an enzyme like a sidekick and alter its structure, or it may help carry glucose through the cell membrane as part of its function.*

Our health simply cannot wait for the exact role of each element to be discovered.

If a cell exhibits the complete chemistry that should occur, and the food which has been ingested was grown in sea water or on sea solid fertilized soil, the cell will most probably be just as resistant to disease as the cells of plants and animals are in the sea. If our present diet does not permit us to take in a complete chemistry, then our cells are incomplete and are subject to invasion by foreign organic matter such as bacteria, virus or fungus. What is even more insidious is that, although we may not have a known or diagnosed disease, we may be suffering from the "disease of dilution," characterized by an organism that malfunctions by comparison with its potential. It is always interesting to read the tremendous amount of research that has been done on disease resistance or the effects of medication and note the statistics. One is constantly faced with the fact that a certain percentage responded and a certain percentage did not. The question "why" is prompted when one is faced with the results of these tests and the answer is that the test subjects' chemistry was obviously different, comparatively speaking. If one were to analyze the food that was eaten by the animal and/or human subjects in the experiments, one would find that their food intake varied tremendously in elemental composition and, therefore, nutritional value as a direct result of the chemical imbalances of our soil.

I began my research 35 years ago because I felt that we should put all of the elements back into the soil in the same proportions that they are found in the sea. I felt strongly that the plants should have the opportunity to take up any element they might need. The possibility also exists that a plant may take up certain inorganic elements that, while not critical for its own physiology, are required by animals in an organic form and only plants can perform the necessary transformation.

Experiments indicated that land plants will tolerate from 400 cc to 1,000 cc of sea water to one-third cubic foot of soil. When sea water is dried by evaporation, the remaining sea solids can be administered as regular fertilizer to the land in the amount of 500 to 3,000 pounds per acre. It was also noted that unless serious rain water runoff occurred, this single application would last four to five years. Corn, wheat, oats, barley, bay, fruit trees, all vegetable crops and other plant life were raised on sea water or sea solid treated acreage. The tolerance experiments indicated that the sea can be recycled back to the land masses and the resulting color, disease resistance, taste and production yields were outstanding. A summary of my research findings is presented in this book.

Chapter 4
Sea Energy Technology

Keep one thing forever in view—the truth; and if you do this, though it may seem to lead you away from the opinions of men, it will assuredly conduct you to the throne of God.

—Horace Mann

It is known that the structure and functions of all plants are a matter of chemistry. When the seed is planted, its first cell divisions are dependent upon the outside environment for moisture only. It is after the seed sprouts that the chemical building of the plant becomes totally dependent on the outside environment. The plant can be built only of nutritional elements that are available to it through the soil. All of these elements must be in an inorganic state, whether suspended or dissolved in water, before they can be utilized by plants. If all of the required elements either are not available in the soil, or not available in the inorganic state, the mature plant will manifest a different physical chemistry from its ideal potential chemical state.

The principle of hydroponic farming is based upon the knowledge that essential elements must be supplied to the growing plant in the form of various types of dissolved compounds present in the water supplied to the plant roots. Generally, no soil is used with hydroponic farming so that growing plants are supported for climbing, where applicable, by mechanical means such as wires and frames. Crushed gravel or other inert granular material is normally employed to provide a foundation for the plant's root system and feeding is accomplished by flooding the roots several times each day.

Heretofore, part of the problem in hydroponics has been the difficulty in determining just what elements were actually essential for growth of a particular plant species. Some 60 elements have now been positively identified in plants with more than one-third identified as essential for complete plant or animal nutrition. Still more elements are on the probable list. About 95 percent of the dry weight of a green plant is composed of the four energy elements of carbon, hydrogen, oxygen and nitrogen. Much of the remaining weight consists of the major ash elements, also called macro nutrients, which include phosphorus, potassium, calcium, magnesium, silicon, sodium, sulfur, boron and chlorine. Less than one percent of the total dry weight is accounted for by trace elements, or micronutrients. Although present in very small quantities, these micronutrients are just as essential to growth as the elements composing the greater portion of the plant's dry weight. Accordingly, conventional hydroponic nutrient solutions have usually contained only the macro nutrients although sometimes traces of iron, zinc and copper have been included. Recently the extreme importance of including trace elements in common fertilizers for soil farming has been realized. Though they play a small role quantitatively in the chemical structure of living plant organisms, many trace elements have already been found to be essential to the growth of certain crops.

It has now been discovered that the most effective nourishment can be provided to hydroponically grown plants by supplying all necessary elements in a predetermined ratio in the form of inorganic salts dissolved in water. Surprisingly, this is accomplished by making up a nutrient solution of those elements in substantially the same ratio with each other as the particular elements are found in sea water. Thus, all of the essential nutrients can be supplied in the proper proportions by using a single solution consisting of diluted sea water plus nitrogen. Preferably, these solutions are obtained by dissolving complete sea solids in fresh water to form dilute solutions containing approximately 1,000 to 8,000 parts per million of sea solids.

These results are all the more surprising in view of comparative experiments which have been made using solutions containing equivalent amounts of sodium chloride only. It was observed that, while dissolved sodium chloride solutions are definitely toxic, solutions of complete sea solids containing the same quantity of dissolved sodium chloride can be used beneficially as a nutrient solution for the growing plants. In other words, sodium chloride is necessary for a complete chemical balance.

Generally speaking, any type of multicellular plant life can be grown hydroponically as long as the water contains dissolved sea solids. The sea solids can be obtained in abundant supply from naturally occurring sources where the sea water has become trapped in shallow coastal areas and dried to completeness, or it can be manufactured directly by evaporating the sea water. It is essential that the entire mineral content be retained in the drying process so the final product contains all the inorganic elements originally present in the sea water, including the original quantity of sodium chloride.

The hydroponic system of food production can be applied most beneficially in the following areas:

1. High income crops such as fresh tomatoes have been outstandingly profitable.
2. Where high levels of quality control are desired as in the baby food industry.

3. In geographic areas of the world where top soil is seriously depleted, absent, naturally rocky or sandy in texture.
4. Plants grown hydroponically use only two percent of the amount of water needed for comparable soil production so that the system is particularly amenable to arid or semi-arid areas of the world.

The research reported in this chapter is in the nature of pilot projects. A tremendous amount of further research still needs to be done, including repeating those items reported herein, to render conclusive the appealing results and provocative trends that have been indicated to date. In my early work, sea water secured from all the oceans of the world with cooperation of the United States Navy and shipped by tank car to Cincinnati, Ohio was used as the basis for the experiments.

In an attempt to develop a stable chemistry in plants and animals, I considered the idea of recycling the sea by using sea water or sea solids as a balanced fertilizer. In the process of developing my plans, I observed an interesting item in the literature, to wit: the quantitative analysis of the elements in the blood has essentially the same profile as the quantitative analysis of elements found in sea water, including the presence of large amounts of sodium chloride. This fact has been surprising to many scientists to whom I have talked who are not primarily engaged in the field of human physiology. The amount of sodium chloride contained in sea water will cause many to question its use as fertilizer because it is well known that salt has been used throughout history as a way to kill plant life on land. As shown earlier, the secret lies in the use of proper quantities of sodium chloride in proper balance with other nutrients, a balance that characterizes sea water and sea salts.

In 1940, a plot containing four 12-foot high peach trees located approximately 20 feet from one another was selected to begin the experimental process of determining the effects of our fertilization process and resulting resistance to disease. The first and third peach trees were designated for experimental tests and were treated with 600 cc of sea water per square foot from the base of the trees to the edge of the foliage to cover the main areas of nutrition. The second and fourth trees were designated the control group and did not receive application. We made the initial application of fertilizer in March before the trees started to bud and around the first of May all four trees were sprayed with "Curly Leaf" virus. The experimental trees remained free of the virus and enjoyed normal fruit yields. The control trees both contracted "Curly Leaf" virus and their peach yield was sharply reduced from the norm. The observation period for the test lasted three years although spraying with the virus took place only in the first year. The control trees contracted "Curly Leaf" each year and finally died while the experimental trees retained resistance throughout the three year test period and provided normal yields each year.

In the same year, turnips were planted in a plot of soil designated half control and half experimental. The experimental section of the plot was fertilized with 600 cc of sea water per square foot of soil after a staphylococcus bacteria associated with "center rot" in turnips had been mixed in the soil of the entire plot. After the turnips had sprouted and the leaves appeared above the soil line, the leaves of both the control and experimental turnips were sprayed with the same bacteria. All of the experimental turnips grew to normal, healthy turnips without evidence of "center rot" while the control turnips contracted staphylococcus "center rot" and died.

It was next decided to grow tomatoes hydroponically in a controlled diet environment for which the following system was used. A box measuring 100 feet by 3 feet by 8 inches, constructed out of cement, was filled with sterilized marbles about three-eighths of an inch in size. The tomatoes were planted in tissue paper a foot apart in the hydroponic beds. A nutrient solution, which was stored in a tank, was flooded into the hydroponic beds, drawn back out and returned to the holding tank three times each day. After the tomato plants had sprouted, their root structure adhered to the marbles, the foliage was tied up and they were pruned to one stalk. The experimental hydroponic bed received a 112 pounds sea solids to 5,000 gallons water solution mixture while the control bed used a traditional hydroponic solution. Both beds were flooded three times daily. Tobacco Mosaic Virus, also lethal to tomato plants, was selected as the exposure disease and all plants were sprayed. As a result, the experimental plants did not contract the disease while all the control tomato plants died of the Tobacco Mosaic Virus.

Hydroponic experiments were conducted both in greenhouse and outdoor settings and the experiment was later repeated in Fort Myers, Florida where the disease incidence is extremely high. Experimental tomatoes grown in Florida during the Autumn of 1970 were never sprayed with insecticide or fungicide and still remained disease free. In all cases the taste was superior and pollination as well as resulting production yields were excellent on the experimental crops. Tomato experiments were conducted in gardens in Northern Illinois during 1954 and 1955. Here the experimental plots were fertilized with 2,200 pounds of sea solids while the control plots were again administered the traditional fertilizing applications. The control plots indicated heavy blight from fungus; the experimental tomatoes that had been fertilized with sea solids were blight free.

Turnips were also planted in the same garden and the patterned results continued. Fifty percent of the turnips in the controlled, untreated plot contracted "etcetera" disease; experimental turnips that were planted in sea solid fertilized soil were disease free.

During 1958 an experiment was conducted, again with tomatoes, in a greenhouse in Skokie, Illinois using soil placed in four cement boxes raised three feet off the ground. The first box was fertilized with an equivalent of 550 pounds of sea solids per acre of soil, the second with the equivalent of 1,100 pounds per acre, the third with the equivalent of 200 pounds per acre. The fourth box was planted as usual by the greenhouse grower for observation by the team. Sample produce from each of the four boxes was taken to the Laboratory of Vitamin Technology, Chicago, Illinois for analysis by Dr. Lawrence Rosner, Lab Director. Table I below details the results of testing.

The moisture content of the tomatoes grown in the 2,000 pounds of sea solids per acre equivalent box was

Table II
Assays:
(1) Moisture   AOAC method
(2) Carotene   AOAC chromatographic method.
(3) Ash   AOAC method.

Results:	Moisture%	Whole Weight*	Dry Basis*	Ash %
Control	90.45	13,000	136,000	.84
Experimental	86.85	18,300	139,000	1.36

*Weight comparisons are given in International Units (IU) of
Vitamin A activity per 100 grams.

reduced over the control box by a slight percentage while vitamin C content increased nearly 25 percent. The specific gravity in the experimental plots also showed a significant increase over the control tomatoes.

In 1957 carrots were tested in Glen Ellyn, Illinois. The experimental plot was prepared with 2,200 pounds of complete sea solids that were worked into the top four to seven inches of soil. Carrots were then planted in both the experimental and control sections. The produce was again analyzed by the Laboratory of Vitamin Technology in Chicago under the direction of Dr. Rosner.

Table III			
Assays:			
(1) Moisture	AOAC method		
(2) Carotene	AOAC chromatographic method		
Results:	Moisture%	Carotene*	Moisture Content%
Control	78.4	19,800	89.9
Experimental	77.6	23,400	83.3

*International Units of Vitamin A activity per 100 grams.

According to the results illustrated in the foregoing Table II, the moisture content of the experimental group was significantly lower than the control carrots. In addition, the whole weight Vitamin A (carotene) analysis showed a greater concentration in International Units by up to 40 percent in the experimental carrots. Dry basis Vitamin A also showed an advantage and the percent of ash ran approximately 60 percent greater in the experimental than in the control. (Ash consists of the weight of the elements after all organic material has been burned out of the sample.)

The carrot experiment was repeated in 1958 and Table III below shows those results including Vitamin A increase and concurrent decrease in moisture content in the experimental crop when compared with the control.

During 1957 a vineyard was divided into control and experimental sections. The experimental grapes were fertilized with 1,100 pounds of complete sea solids and the grapes were taken to the American Research and Testing Laboratories, Chicago. The report, signed by Paul W. Stokesberry, Director, on September 13, 1957 showed an analysis of two samples of grapes to determine total sugars. The control grapes had 13.60 percent total sugars in the juice while the experimental had a significantly higher 16.87 percentage. Whole grape analysis also showed a higher percentage of sugars in the experimental crop; 14.21 percent compared to 11.45 percent for the control grapes.

In 1954 we decided to conduct large field experiments with the use of sea solids as fertilizers. The purpose was twofold: sometimes small experimental plots receive care that differs from larger field experiments and, in addition, we wanted to grow enough oats, corn and soybeans to conduct feeding experiments with animals. The large scale experiments were conducted at Ray Heine and Sons Farms, located on the southwest corner of Rutland Township, 11 miles west of Elgin at the intersection of Illinois State Highway 47 and U.S. Highway 20. The following describes soil experiments in 1954 and subsequent feeding experiments with pigs and chickens conducted in 1955.

Complete sea solids in 1,500 pounds per acre quantity were ground up and applied to an experimental plot measuring 7.5 by 91 feet in a field where the growing corn was four inches high. The results from the above experiments are as follows.

1. Corn Experiments:
   a. Sea solids were in no way detrimental to the growth of the corn.
   b. Uniformity of growth:
      —experimental corn: substantially free of nubbins, uniformly high stalks
      —control corn: usual distribution of nubbins, normal variation in size of stalk.
   c. Yields:
      —experimental ears: 1.5 inches longer on the average than control ears, 3/8 inch larger in diameter on the average than control ears.
   d. The experimental plot yielded four more bushels per acre than the control plot.
2. In a garden experiment, complete sea solids were applied to a 10 by 20 foot plot and worked into the soil before planting radishes, beans, peas, carrots and lettuce. The same plantings were made in a control plot not fertilized with sea solids. All vegetables grown in the experimental plot had a superior taste to those grown in the control plot and the leaf lettuce of the experimental area permitted four cuttings compared with two cuttings of control lettuce.
3. During the next growing season, which followed the preliminary experiments outlined above, the following larger scale field experiments were conducted.
   a. Oats
      April 19-24: Sea solids ground in burr mill to a very fine texture were applied to soil using an International Harvester Ten-Foot Fertilizer Spreader. The 2,200 pounds per acre of sea solids were spread over 10 acres of a 19-acre field, leaving a nine-acre portion of the field untreated. The sea solids were worked into the top four to seven inches of soil using a twelve-foot field cultivator and Bonda oats were broadcast and disced in the complete 19-acre field. Heavy rain fell intermittently through June 4th.

May 3: Observed oats were coming up—control oats appeared to be taller than experimental.

May 7: Control oats were 1 to 1.5 inches taller than experimental.

June 7: Oats in both plots approximately 9 inches high.

June 10: Color difference observed and the exact line where fertilization stopped was apparent through the center of the field. Experimental oats had a much darker green color. Rabbits and grasshoppers were observed to exhibit a marked preference for oats in experimental plot.

June 13: Cows being driven down the road preferred grass at the edge of experimental plot.

June 14: Color difference of oats is more pronounced.

June 18: Oats headed out with experimental oats more advanced.

July 21: Oats in experimental plot are ready for cutting.

July 24: Oats in both plots were cut; experimental oats were found to have less rust.

Yield — Oats:

Control Plot: 38 bushels per acre.

Experimental Plot: 45 bushels per acre.

b. Corn

May 25-30: Manure was applied to 30 acres of a 40 acre field on May 25. In May 2,200 pounds of sea solids per acre were applied in the manner outlined above to a 10-acre plot, retaining the remaining 30 acres as the control plot.

June 8-9: Pioneer corn planted in entire field along with 50 to 80 pounds per acre of a nitrogenous fertilizer material (Commercial 2-12-12 Fertilizer).

June 14: Corn showed above the ground and no apparent difference between experimental and control was noted.

July 22: Corn was observed to be tasseling.

August 1: Tasseling of control corn was further

advanced than the experimental.

August 23: Corn in both plots was observed to be the same height and color. Each hill of corn on 4.9-acre portions of experimental and control plots was inspect ed for galls (smut). Control corn had 384 percent more observable galls than experimental corn.

Yield — Corn

Control Plot: 75 bushels per acre.

Experimental Plot: 88 bushels per acre.

4. During the next season, 306-day-old New Hampshire chickens were obtained for feeding experiments using oats and corn grown on the sea solid fertilized soil during the previous season. The control group of 153 chicks was fed commercial concentrate plus a feed mixture of two parts corn and one part oats grown on control plots. The 153 experimental chicks were fed the same mixture as the control group with the exception that the corn and oats used were grown on the experimental plots and, thus, were fertilized with sea solids. The following results were obtained.

Roosters:	Control Group*	Experimental Group*
At 4 months	42	60
At 6 months	106	128
At 2 years	135	152

(*average weight in ounces.)

Hens:		
At 6 months	80	104
At 2 years	96	114
Time of laying	5 months, 3 weeks	5 months
Eggs (weight/dozen)	19 to 23 ounces	23 ounces
Eggs (post 7 months)	24 ounces	28 ounces

Entire Group (Roosters and Hens):

Average feed consumed per pound of of weight gained (in lbs.)	3.0	1.89
Mortality	3	0

Diseased		
Worms	Yes	No
Nervous Condition	Yes	No
Leg Disjointing	Yes	No
Size	Varied	Uniform

5. One sow and six pigs raised on corn and oats grown on land fertilized with complete sea solids were unusually uniform in size, showed no tendency to "root" and were easily contained in a small fenced area. When they reached approximately 180 pounds, they were taken off this feed and given control corn and oats. They immediately began extensive rooting and, by the end of the third day, they were extremely nervous and broke out of the pen on two occasions. On the fourth day they were put back on sea solids grown feed and were calm by evening. Thereafter, they were easily contained in the pen and, again, showed very little rooting tendency.

Table IV

Samples:	Ash Weight in % Solids		% of Increase
	Control	Experimental	
Onions, Bulb	13.6	14.2	4.4
Oats	87.7	87.8	0.1
Sweet Potatoes	28.8	31.2	8.3
Tomatoes	4.8	5.7	18.7
Soy Beans	73.9	84.7	14.6
Corn	73.1	74.4	1.7

Table IV shows the ash weight determination of field oats, corn and soybeans, tomatoes, sweet potatoes and onions that were raised on a garden plot at Elmhurst College, Elmhurst, Illinois in 1955 with the equivalence of 2,200 pounds of sea solids per acre on half the garden. The other half was fertilized normally.

Additional feeding experiments were conducted at the Stritch School of Medicine, Loyola University, Chicago, Illinois using the experimental and control oats, corn and soybeans grown during 1954 at the Ray Heine and Sons Farms as described above. I want to emphasize that these feeding experiments and the results are only preliminary and it must be kept in mind that the mice, rabbits and rats used in these feeding experiments have a different physiology than human beings. The results are not definite but merely indicate an interesting trend and further research should be done to further document the findings. To date, I have not had the research funds necessary to repeat the large scale field experiments that would be required to produce the volume of food to duplicate these feeding experiments. In no way, however, do I suggest that the same results would occur in a human being due to the preliminary stage of research.

The following is drawn from my laboratory notes covering the Ray Heine and Loyola research experiments.

Started feeding mice both experimental and control food that was raised on the Ray Heine and Sons Farm. The experimental food had been raised on soil fertilized with 2,200 pounds complete sea solids. The control food was the same as the experimental with the exception that it was not fertilized with complete sea solids. The food consisted of a combination of one part soybeans, two parts oats, four parts corn, balanced food proteins, carbohydrates and fats for mammals.

C_3H mice were obtained for this feeding experiment. This strain of mice has been bred so all the females develop breast cancer which causes their demise. The mice were two months of age when received and started on the feeding experiments. The life expectancy of this strain for females is no more than nine months which includes the production of two or three litters. The experimental and control groups both consisted of 200 C_3H mice and those fed on control food were all dead within eight months, seven days. The experimental mice that were fed food grown on the sea solids fertilized soil lived until they were sacrificed at 16 months; definitive examination revealed no cancerous tissue. The experimental group produced ten litters compared to the usual two to three litters and none developed breast cancer.

Spraque Dolly rats were obtained and were divided into groups of 25 control and 25 experimental. The control rats were fed controlled food while the experimental rats received the sea solids fertilized food. Both the control and experimental groups were injected with cancer (Jensen Carcino-Sarcoma) which has been shown to be a 100 percent killer. All of the rats fed on the control diet died within 21 days of cancer. Nine of the rats that were fed the experimental diet died of cancer within 40 days; sixteen lived five months until they were sacrificed; there were no cancer "takes" in the sixteen out of twenty-five survivors that were fed experimental food.

One hundred and twelve rats were fed on experimental food for a six week period. Then half of the rats were sacrificed and the thymus gland was removed and implanted in the remaining 56 experimental rats. (The experimental group then contained the equivalence of a double thymus gland.) Jensen Carcino-Sarcoma was then injected in all 56 control and 56 experimental rats with the result that all 56 control rats were dead within 23 days. Of the experimental rats, two apparently had a cancer "take" but it was absorbed and disappeared. Four of the 56 experimental rats died of cancer and the remaining 52 were sacrificed 90 days after their original cancer injection. No cancerous tissue was found in these 52 experimental rats.

Again, let me repeat that this was a feeding experiment that was conducted on rodents and not human beings. Although it indicates a possible trend in disease resistance due to food grown on sea solids fertilized soil, the part played by the double thymus gland must also be determined before conclusions can be drawn. Consequently, this experiment needs to be repeated by a wide spread of teams of investigators to determine ramifications.

In the next experiments, twenty-four rabbits were obtained.

Twelve were designated experimental and fed on food grown on sea solids while the remaining twelve were labeled control and fed accordingly. All of the rabbits were given a high cholesterol diet for six months which produces hardening of the arteries. The control group did develop hardening of the arteries and all had died within ten months. The experimental group did not exhibit hardening of the arteries.

A breed of rats that developed disease of the eye was obtained. The 10 that were put on experimental food showed no deterioration of the eyes and bred five litters. Those on the control food diet all died secondarily of eye disease.

Hay was grown in Lennox, Massachusetts on soil fertilized with 2,200 pounds of complete sea solids. Corn and oats grown in Ohio and Illinois on soil treated with complete sea solids were also obtained and fed by a dairyman to pregnant cows. One of the problems previously experienced by the dairyman was that his newborn calves from these pure bred cattle had difficulty standing in order to nurse when they were first born. They often had to be held for their first nursings and were often not uniform in size. However, when calves were born from the cows that had been on food grown on complete sea solids fertilized soil, all of the calves were immediately able to stand up to nurse and were uniform in size.

In 1970 an experiment was conducted in southern Wisconsin, the report of which follows:

A 40-acre field on which corn had been grown for the preceding nine years was treated in 1969. Although a portion of the field required three tons of lime per acre, we applied four tons. In the spring of 1970, 110 pounds of anhydrous ammonia was added to the entire field followed

by an application of sea solids to 14 of the acres which con-
stituted the experimental segment. The rows were 80 rods
long, with each plot about an acre, or 10 rows, in size. Two
hundred pounds of sea solids were placed on the first acre.
Each of the remaining acres had an additional 100 pounds
poured on it than the one before, so that the last acre had
been enriched with 1,500 pounds of solids.

Kings Cross Corn PX-610 was planted along with 150
pounds per acre of 6-24-24 starting fertilizer. Germination
of all corn was excellent. Throughout the growing season,
the crop showed a stair step effect in growth with corn on
the 1,500 pounds of sea solids per acre showing the best
advancement. All plots except for that one exhibited corn
blight, marked by fallen stocks, and a difference could be
easily noted in the effect of the blight starting with the 400-
pound plot and diminishing with the 1,500-pound acre.

The corn was harvested in each test plot on November
7, 1970 by one round with picker to determine how well it
shucked, and to determine blight damage to cobs and kernels. The remainder was combined. The 1,500-pound corn
yielded 154 bushels per acre, while the untreated acre yielded 115. The yield increased as the sea solids increased.
Weight of the 1,500-pound corn was 57.5 pounds per
bushel compared to 53.5 pounds per bushel for the untreated acre. The 1,500-pound corn consisted of 20 percent
moisture while the untreated consisted of 25 percent. The
treated corn leaves were much greener at harvest even
though the corn was less moist.

There was some evidence of corn blight on the 1,500-
pound corn leaves but it did not affect the ears. The cobs
shelled out with complete, whole kernels and the cob was
solid. The untreated and low application (100 and 200
pound) corn suffered ears with rot at the ends of the cob.

This same seed corn (described above) and fertilizer
treatment was planted in another field which had been used
the previous year to grow alfalfa. Manure was spread and
the alfalfa field was plowed under to prepare for the seed

corn planting. Although no sea solids were administered, the yield on this plot was 130 bushels per acre. However, the best corn yield, weight and moisture on the 3,000-acre farm was the product corn received from the 1500 pounds of sea solids acre.

Steers weighing 1,100 pounds that had been fed regular corn were fed on corn grown on the (above described) 1,500 pounds per acre plot. The steers were fed on up to 1,400 pounds using one-third less corn than was previously required with regular corn and they appeared to be in very good condition.

Field tests have been conducted in South Dakota, Wisconsin, Illinois, Ohio, Pennsylvania, Massachusetts and in Florida. All field results were essentially the same no matter what type soil was used. Production was the same or greater on that soil which was fertilized with sea water or with complete sea solids. Animals that were fed on field crops preferred the experimental food. Experimental crops were consistently more disease resistant than the control. Garden vegetables and fruits were superior in taste. Onions, tomatoes, potatoes, sweet potatoes, apples and peaches were outstanding in taste and onions could almost be eaten like apples. People who ate the garden produce said that in spite of the superior taste they did not seem to eat as much of any of the vegetables as they normally consumed. They became full or their appetite was satisfied on less food. The most dramatic results in plants was in the second and third generations of those that were raised on complete sea solids fertilized soil. The feeding experiments involving pregnant animals fed on sea solids fertilized crops produced young that were very uniform in size and all of the offspring seemed to respond in dramatic fashion to this balanced chemical diet.

Fortunately for our potential future health, changes also occurred in the adult animals when they switched to food grown on the sea solids treated soil.

There is a very real and pressing need to develop an experimental animal that has consistent chemistry. I suggest we raise "mini-pigs" on food grown on sea solids fertilized soil. Pig physiology is much closer to human physiology than normal laboratory animals such as dogs, rats, mice, guinea pigs and rabbits. If we develop this breed of animal with muscle tissue, nerves, lungs, heart, kidneys, etc., that are characterized by constant body chemistry, we then will have an animal that ideally lends itself to very precise research. This would aid existing efforts in testing drugs, antibiotics and disease resistance in general.

Of equal importance and equal urgency is the need to conduct experiments to see what the long-term effects food grown on sea solids will have on human beings. We must recycle the sea—*and now*—for our health and the health of future generations.

Chapter 5
Pictorial Interlude

Many persons reading this book will want to experiment further with sea salt technology in their gardens or with houseplants. Farmers may even be interested in spreading sea solids over their acreage. We certainly hope so. Provoking serious interest has been the purpose of this book. There are, however, a number of variables and ramifications involved with the patented process of sea salt application so the do-it-yourself gardener may need further assistance. It is not recommended that you immediately run to the ocean to get a bucket of water for your plants.

Sea water is excellent fertilizer, but it must be applied properly. All the variables in soils and plant types must be taken into consideration. It took Dr. Murray 40 years to perfect his process and cope with all the variables. His organization established methodologies and assembled a crew of technical advisors to work with prospective sea salt experimenters and patent licensees.

This technology has been put on hold by Dr. Murray's death, and asks readers to pick up where these paragraphs leave off. The reasons are contained in the remaining chapters of Dr. Murray's book, but in this chapter, the pictures tell the story.

Dr. Murray and one of his sea salt nourished, hydroponically grown tomato plants.

A remarkable comparison was made between experimental and control hogs. The experimental animal is more than twice the size of the control.

Animal life in the sea is far healthier than similar life on land. Tissue samples from an adult walrus compare with those of a baby walrus.

Trout of the same species—the larger is a sea trout, the smaller a fresh-water trout. The difference in size is not usually so pronounced, but in health the difference is tremendous.

Sea solids were spread on farm acreage many times during early exper-iments. The line clearly marks experimental and control land.

All plant life responds favorably to sea salt nutrition. Here are mature asparagus plants. On the left is the control plant, obviously less bountiful than the experimental.

Results of sea salt nutrition experiments with apple trees: the smaller is the control.

Hydroponics is growing plants without soil. Gravel beds are flooded with sea salt nutrients twice each day. Growth rate, general health and yield are far superior to other methods. This photo shows bed preparation.

Young cucumber plants shortly after planting.

Cucumber plant growth.

Cucumbers nourished by sea solids mature rapidly and produce bumper crops, as seen here.

Hydroponically grown tomato crop.

Tomatoes experience the same luxuriant growth patterns as cucumbers. Note the abundance of the crop.

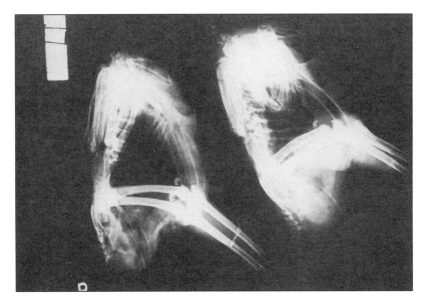

Animals fed crops nourished with sea solids also do better. This X-ray photo of chickens shows the experimental birds with solid leg joints, whereas the control birds had the disjointing characteristic of poor nutrition.

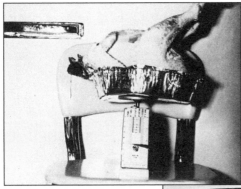

Weight comparison of dressed experimental (right) and control (above) birds.

Chapter 6
Recycle the Sea
for Better Human Health

We cannot impose our wills on nature unless we first ascertain what her will is. Working without regard to law brings nothing but failure; working with law enables us to do what seemed at first impossible.
—Ralph Tyler Flewelling

Finding the solution to the health and aging problems which beset man has always been one of life's great challenges. Curiously, some of man's most highly developed technological civilizations failed miserably in this quest as we are failing today, while certain "primitive" groups in isolated locations enjoy excellent health and unsurpassed longevity records. Even if man never suffered from disease, the mysterious process of aging would constitute more than sufficient challenge to science. Man has fought to preserve the vigor of youth against the ravages of age since thinking began. History tells us Ponce de Leon sailed across the Atlantic in search of the fabled fountain of youth. How ironic that he most likely was floating on the surface of the only realistic means of staying young and healthy!

The waters of the oceans hold the perfect balance of those essential elements required as food for the complex cell groups that make up our bodies. This is my thesis—now for the proofs.

When I was going to school at the University of Cincinnati in 1932, I attempted to induce cancer into a toad, but was astonished to note that the amphibian seemed to have a natural immunity. This laboratory incident precipitated the beginning of a lifelong search for an explanation. In 1966, I fed crops grown with recycled seawater to various farm animals and obtained remarkable health and growth results which confirmed my theories. As I write these lines, I am earnestly trying to get stubborn establishment thinking to wake up to the very real dangers facing mankind and life on this planet. Disease and famine pose very real threats, not just to parts of the world that seem distant and unreal to us in America, but to our very own bustling civilization.

It is extremely interesting to carefully examine the biological activity in the sea. A cubic foot of ocean water sustains many more times the number of living organisms, plants and animals than does the equivalent amount of soil. Sea water is literally alive, especially where the temperature of the water is warm.

Of special interest is the fact that the aging process does not appear to occur in the sea. A comparison between the cells of a huge, adult whale with cells taken from a newly born whale will show no evidence of the chemical changes observed when comparing cells of adult and newborn land mammals. There are some denizens of the sea that apparently never cease growing. One need only compare the size between land turtles and sea turtles to realize the tremendous difference. Some zoologists would claim they are turtles of different species, but I disagree. I am convinced that the difference in size and longevity is due to the complete, balanced chemistry provided by the sea environment.

There is no chronic disease to be found among fish and animal life in the sea that compares to those on land.

Science is aware that nearly all individual cells in an animal body are replaced during the process of cell division. In man, for example, most of the cells are replaced within about 18 months. If the requirements for certain elements are not supplied by the food ingested as cell division occurs, dilution becomes apparent until these critical elements are nonexistent in the organism. This shortage of essential elements does not occur in the sea. Why aren't these vital elements in our food?

You and I can't tell by looking at a carrot or tomato that the essential elements are missing, however molecular biologists find substantial differences between two vegetables from the same species. A plant can grow to maturity, and yet make dangerous substitutions of elements in its structure due to its chemical attempts to compensate for an imbalance of the proper elements in the soil. If our cells in turn must compensate for the dilution, or lack of elements, then they lose their resistance to disease. Remember, our bodies are host to an enormous number of microbes that eagerly pounce when the slightest breakdown in cell function occurs.

To me it is only logical that the cause of our frightening increases in chronic disease and the sorrowful process of aging is the absence of a complete, balanced physiological chemistry.

If the necessary elements are not found in our food, where are they? Certainly nature has provided them. The answer is that they have departed from our soils due to continuous taking of crops and the process of erosion. Most crops utilize an average of 40 elements from the soil. In no case do fertilizers add more than 12 and most commercial fertilizers add a maximum of six elements.

The singular most devastating source of depletion of soil is water leaching. Even on relatively level land tremendous leaching occurs and has been taking place for thousands and thousands of years. Ultimately, the various leached elements, because they are in water solution, flow down to the sea.

I once stood near the mouth of the Mississippi River and watched the muddy outpouring into the gulf. Within 24 hours the mighty Mississippi deposits topsoil equivalent to a 120-acre farm into the sea. It is not enough to merely control gully erosion which is widely recognized as the major problem faced by soil scientists today. If all erosion were halted this instant, we would still have soil that is seriously depleted of the balance of elements required by body cells.

For countless centuries the vital elements have eroded off into the sea. What state are they in while mixed with our vast oceans? Analysis of sea water shows a constant proportional balance of all the water-soluble elements. If an excessive amount of any one element flows in due to erosion, it drops out to the bottom of the ocean. Three and one half percent, by weight, of the sea water is composed of sea salts or sea solids. Ocean water may taste salty, similar to our table salt, but careful examination reveals that sea solids are darker in color and chemical analysis shows that all the elements in the atomic table are present with the possible exception of some of the gases.

I have used these sea solids as plant food in experiments to prove that these elements in perfect balance will grow chemically perfect plants. Note that I did not try to synthesize anything, but merely took what nature already offered.

My first experiments were conducted in 1938. Since then I've carried out literally hundreds of experiments involving feeding plants nothing except sea solids mixed with tap water and a minor but fertilizing amount of a water-soluble nitrogenous material such as ammonium nitrate, sodium nitrate, potassium nitrate, calcium nitrate and the like, which form nitrate ions when dissolved in aqueous solution. Invariably the result has been the same—healthier, more productive crops. Early in the experimental game I learned that hydroponics, which is feeding nutriments to plants without soils, gave me better control over the plant diet. Dried, natural sea solids were dissolved in plain water, using approximately 112 pounds of sea salts to 10,000 gallons of water, which is a damned economical mix. The only nutrition my experimental crops received were the sea solids in solution which bathed their roots a few times each day. The plants flourished as no plants have flourished in the modern day of fertilized soils. The contrast in the experimental crops with the control crops grown by normal commercial methods was truly exciting. The taste difference was very significant, especially in tomatoes and carrots. The production rate was considerably higher and the resistance to disease was apparent.

The second line of experimentation was to put these evaporated sea solids directly on the soil as fertilizer. We actually used up as much as 3,000 pounds per acre—and I know eyebrows are raising now!

Many people are familiar with the story of how the Romans destroyed the land around ancient Carthage with salt, and it's true that our table salt in the amounts we used as fertilizer would be harmful to the plants, perhaps would kill them. But in the presence of the other elements as found in sea water, sodium and chloride are not toxic to plants. Actually salt maybe necessary for the absorption of the heavier elements. It is known that a saline solution will pick up a greater quantity and variety of elements than ordinary water solution. The elements in soil must be dissolved in water since this is the only way they can be absorbed into the root system of the plant.

We planted fields side by side so that one experimental plot used sea solids mixed into the soil as fertilizer and one control plot used the best commercial methods available. The results were similar to those with the hydroponic system. The sea solid fertilized crops grew faster, were healthier and produced a far greater yield. The colors of the plants also differed and a taste difference was obvious. Animals, both wild and domestic, had no trouble determining which was better for them to eat and one walk through a field of oats showed us a glimpse of animal heaven. Rabbits and mice scurried everywhere, yet the minute we stepped into the control area where standard fertilizers had been used, it was almost lifeless so far as the animals were concerned. We decided to play a game and put a little tape around some green stalks of field corn to identify them as having come from our experimental field. We mixed the experimental with the control stalks and placed them in the feed lot for cattle and sheep. We watched closely as the animals munched away. It was immediately apparent which stalks they preferred because after once sampling an experimental stalk, the animals would nuzzle and burrow in the pile to find another stalk, ignoring the control stalks until they had no other choice.

To further prove that animal instinct knows best, we treated a section of a clover field covering an area of about 100 square feet with the sea solids. When the clover grew to around six inches, sheep were let out to graze. They walked and grazed until they came to the treated spot, then ate until the clover within the treated area was nubbed to the ground.

Feeding experiments with steers showed that they had greater weight gain while eating less of the experimental feed. Farmers ought to appreciate that. Although detailed accounts of our experiments have been covered, I want to share one further animal feeding test. We used 306 freshly hatched chicks and designated 153 control and 153 experimental. The experimental group was fed a commercial concentrate and oats along with corn and soybeans grown on sea salts treated soil. The control chicks were fed the same diet with the exception that all feed was grown on nontreated soil. At the end of six months, the experimental roosters weighed a full one and one-half pounds more than the control group. Experimental hens laid eggs for the first time fully one month earlier than the control hens and also exhibited a phenomenon amazing to anyone who is familiar with laying hens—not a single experimental hen laid a pullet size or small egg! All of the experimental eggs were of firm shell and large size. During one complete year of careful observation, the experimental chickens exhibited perfect health. They were free from disease, and furthermore they remained calm when approached by men. The control chickens were nervous when approached by the flock tender, they exhibited disease such as slipped tendons and worms and several died of unknown causes. None of the experimental chickens died.

Similar advantages to food grown with sea salts were seen in experiments with laboratory rats. The control rats showed less weight gain per pound of food and sustained definite eye disease. The experimental rats, on the other hand, exhibited sleek coats, were apparently immune to the eye disease that afflicted the others and showed a markedly uniform weight gain on less food. We then conducted a similar experiment with mice bred to develop breast cancer and the experimental mice failed to develop cancer and lived significantly longer.

Wow! You might exclaim, why not sprinkle sea salts on our foods and get healthy? It simply doesn't work that way. Anyone with a cursory knowledge of biology knows that humans and other animals cannot obtain any benefits from the elements unless they have been hooked up with a carbon atom by the green plants. It is obvious to me that this is the explicit role of plant life on earth, i.e., to convert inorganic elements to organic compounds which can be utilized by animal life. Table salt is the only food we eat that is inorganic and, frankly, it isn't very good for us.

Sea energy agriculture, which is growing foods with sea solids as fertilizer, provides a means for improving our chemical intake without sacrificing our eating habits. Our meats, vegetables, fruits and cereals would all be adequately balanced with the essential elements simply by growing all crops with sea salt technology.

It has been shown by agronomists that soil may contain a large amount of one particular combination of elements, yet the plants cannot absorb them. The complex molecules of living tissue in plants and animals are made possible by the carbon atom. The linking up process is made possible by the various elements in combinations called catalysts and these catalysts invariably have a critical minor element or "trace element" that apparently serves as the key to their function. The presence or absence of a trace element can be the deciding factor in determining whether a necessary element is absorbed into the plant's root system. The balance of elements must be right in the soil for plants to synthesize their complete chemistry.

Tomatoes serve as an example of the need for this balance. There may be a few individuals who know as much about raising tomatoes as I do, but there's nobody who knows more. Tomato growers know that potassium is a macro element, or an element with a major function in the plant's growth. Potassium is added to the soil in quantity by tomato growers. Yet the tomato itself has only a minor amount of potassium in the mature product. My hydroponic experiments proved conclusively that only a small amount of potassium, as found in its proper balance in sea water, was needed to grow outstanding crops of unusually healthy tomatoes. My point is that it is unnecessary to fertilize heavily with one element or another if an adequate balance of elements can be made available for the plant's use.

Growing staple crops hydroponically in sea water solution has tremendous implications, especially for the starving millions in our world. One super advantage is that plants grown hydroponically require only about one-tenth the water that the same number of plants growing in soil require. The cost of hydroponic facilities becomes negligible when the exceptional productivity is considered. It sometimes burns me up to read what our establishment scientists have to say about the world's food problems—in fact I'm going to digress from my topic again.

When the technical journals stress that the "long term solution to the food crisis is development of new, productive crop hybrids and the spread of modern agricultural technology throughout the developing world," I shudder. The established experts harp on things like "pest control," better management of "fragile soils" (I go for that) and novel ideas for "storing water," but they turn a deaf ear toward sea salt technology which provides all of these things naturally. Of course, there are economic pressures and it is unrealistic to think the large fertilizer companies want to go out of business which is precisely what will happen when sea salt technology takes its rightful place in our "modern agricultural technology."

Aside from economic and productivity implications, what are the implications for man if we are able to restore the chemical balance to our food? We can eliminate illness as we experience it today. I know that to many of you this sounds like a grandiose, unproven claim, but one must remember that we are only beginning to investigate a new agricultural technology.

One of the most exciting prospects is that perfect nutrition could increase man's brain functions far beyond the presently exhibited capacity. Consider the estimates by neurophysiologists who say that we use from one to ten percent of our 10 billion brain cells. The results of a more complete utilization of this particular cell group due to balanced physiological chemistry could be beyond our imagination.

I am convinced that sea salt technology could be the way we humans finally learn to use our heads to solve serious problems. One wonders if things have always been this way? Have land plants, animals and human beings always been as they are today? No one can be certain, but perhaps at one time there really was a Garden of Eden where man's longevity was extended well beyond what it is today. If continents have truly been sloshed around in the sea as cataclysmic geologists tell us happens every now and then, evidently sea salts bathed the land masses and provided the survivors of the cataclysms excellent nutrition.

The latter is all speculation, of course, but it should be possible to locate pockets of preserved, complete soil on earth even today, thereby illustrating the value of balanced elements. Many investigators proclaim they have found these areas where the soil has more complete chemistry, such as the Valley of the Hunzas in the Himalaya Mountains of Asia. In this elevated and isolated place, observers have reported that men and women live to a vigorous 120 years of age, that they are able to procreate and bear children at 100 years or older and that there are no "chronic" diseases.

Other investigators have found a tribe in Northeastern Africa where the individuals possess phenomenal hearing and are in excellent health without chronic diseases. It has been learned that whenever these tribesmen go down to the coast to inhabit a "civilized" environment, which includes eating the food of civilized society, their health deteriorated sharply. It doesn't take long before these super-healthy specimens begin to resemble the rest of us. In fact, I find it amazing that the human is so sturdy after coping with civilization's arrogant disregard for nature for so many thousands of years.

Still another isolated area is found in the valleys of Columbia in South America. Some individuals there claim to be 140 to 160 years of age and are in excellent health.

Further investigation shows that these individuals come from valleys which are completely surrounded by mountains, and therefore minimal leaching or erosion of soil has occurred throughout the centuries.

These differences between peoples should tell us something.

Dr. Eugene H. Payne, an investigator for Parke-Davis and Company, a pharmaceutical firm, published an article in *This Week Magazine* (August 8, 1954) entitled "Medicine's Most Amazing Mystery." He related how he found six areas in South America that seemed "magically" free of cancer, heart disease, malaria, tooth decay, hookworm and insanity. He also stated that hidden "somewhere among them—in the water, the rocks, the soil, the food, or perhaps even in the minds of the people who live there— are six medical secrets so precious that to discover them and put them to scientific use might easily alter the whole course of mankind."

Dr. Payne found in a province of Loga, in Equador, an area covering about 500 square miles where the people exhibited no signs whatsoever of heart disease or circulatory disturbance. There were plenty of diseases such as malaria, dysentery and typhoid, but no cardiac problems. The researcher had visited Loga to find an old friend who had retired to the region because of high blood pressure and heart problems. When the friends met, Dr. Payne learned that the retired man had normal blood pressure and no heart abnormality. However, the researcher's friend reported, whenever he left the area it didn't take him long for his blood pressure to increase and his heart begin to give him trouble. Loga, Equador, at least in the 1950's, was an island of immunity from heart disease.

In 1943 Dr. Payne found another island of immunity. This one was located about 200 miles north of Lima, Peru in a place named Callejina-Huaylas, which is nestled in a 75-mile-long valley high in the Andes Mountains. All the residents, without exception, were completely free from hookworm, an ugly, debilitating and often fatal intestinal parasite which normally flourishes in South America.

In Minas Gerais, Brazil, Dr. Payne found an area of the country where everyone over the age of 15 years had far too many tooth cavities. When he checked the water he found it to be very high in fluorine—the element we advertise as fighting against tooth decay and mix into our drinking water. Because there were feldspar mines nearby, the fluorine was present in unusually high concentrations. Yet, in other areas of Brazil, where Dr. Payne found that fluorine was totally absent from the drinking water, the incidence of dental caries was lower than any reported in the United States. That's curious, isn't it? The fluorine didn't help a bit in the area where the soils had been depleted, and the highly advertised element wasn't needed where nutrition was improved. Although Dr. Payne's research expedition did not thoroughly analyze the foods and the soils of these various peoples, he suggested it should be done, and I definitely agree.

A more recent report on the geographical differences in disease was written in the May, 1971 issue of *The M.D.*, authored by Dr. M.J. Hill and co-workers of the Wright-Fleming Institute, St. Mary's Hospital, London. Their research thoroughly investigated the relationship of cancer of the large bowel to the chemistry and bacteriology of stool specimens. The six locations studied included India, Uganda, Japan, England, Scotland and the United States. The first three nations have a low incidence of large bowel cancer while the latter three have a high incidence of the disease.

Keeping in mind what has already been said about the importance of microbes and the balance of elements, we studied this 1971 report. The low incidence countries, whose diet is low in fat and animal proteins, showed more aerobic bacteria and far fewer gram negative anaerobes than the high incidence countries. The Western countries featured a large consumption of animal fat. Anaerobic bacteria metabolize steroids much better than aerobic bacteria thus a high incidence of cholesterol metabolism was found in the feces of the high incidence countries. The concentration of acid steroids derived from bile salts was seven to 11 times higher in the feces of high incidence countries than in the low incidence countries. Dexycholic acid, a bacterial degradation product of bile salts which is considered carcinogenic or cancer causing, correlates with the higher incidence of colon cancer. The British oncologists concluded that the results "strongly supported the postulate that the geological differences in the incidence of colon carcinoma may be related to dietary habits and that these could operate through their influence on the nature and number of intestinal bacteria."

That report, and this discussion, remind me of the statement by a pediatrician who said that if people think they are really in good health, then they should ask themselves if they can defecate and not need to wipe the residue from the anus. People in animal husbandry know that illness is indicated when animals do not have clean bowel movements. I am convinced that if the chemistry of the human being was as it should be, fecal matter could be defecated in the same manner as horse, dog, or sheep without leaving a residue.

I realize it is harping on the subject to continue in this vein, but it appears the facts must be drilled home or we will continue munching away complacently and paying through the nose for medical treatments. Our accepted notions, played to the hilt by the advertising media, are every bit as sick as our populace. I nearly gagged when I saw the television commercial for Pepto Bismal complacently suggest that indigestion is part and parcel of living, so to avoid the pain, coat the stomach lining. We are deluding ourselves about our progress and condition. For example, in Mexico, a country whose sanitation practices are notoriously poor by American standards and where inoculation is seldom practiced, smallpox is rare. Think of it! The so called "filthy" disease, smallpox, against which we must innoculate ourselves is almost unknown in a country with lower standards of hygiene and sanitation.

And we can't blame it all on the high incidence of animal fat we ingest. The Eskimos and many of the less civilized peoples of Polynesia enjoy diets which are high in animal fat, yet hardening of the arteries is highly unusual among these "primitives." We blame our high incidence of arteriosclerosis on those same fats.

During World War I (circa 1918) with high military standards for physical and mental fitness, 31 percent of all the young Americans called for induction into the armed forces were rejected as unfit. For World War II (circa 1943) the rejection rate was over 50 percent so the standards were lowered to a point below that of 1918. This lowering of standards lowered the rejection rate to 41 percent. During the period between 1948 and 1955, which included the Korean War, the physical and mental standards were lowered even more, yet the rejection rate of young men between ages 18 and 25 climbed to 52 percent. More than half the young men of our nation who were called for military duty were rejected. How can we call ourselves healthy? Any nation with a drug industry flourishing so well as ours certainly cannot claim good health.

Having posed these questions and observations, it is now time to propose an answer: food power from the sea.

Chapter 7
Planned Food Pollution

When a man's science exceedeth his sense,
He perisheth by his ignorance.
                                          —Oriental Proverb

Select 10 people at random and ask them which industry they believe to be America's biggest. About half of them will say "steel or auto," one or two might think chemicals and plastics and the others might vote for oil, or communications such as American Telephone and Telegraph. None of them would be correct. The food industry is far and away America's largest. Yet, for all its size, sales, retail outlets and number of workers, the business of manufacturing food is not a human enterprise. Green plants are the only producers of food on this planet. Human hands and minds are only capable of harvesting, packing, processing and marketing the products which nature alone knows bow to manufacture.

Gathering food has always been and will always be man's first order of business because it is basic to survival. Food getting for the populace has evolved from a nomadic hunting and foraging existence to the modern supermarket with its astonishing assortment of cans, bottles, packages, cartons, bags, boxes and other packaging innovations. The myriad of containers hold an infinite variety of edibles brought in from all parts of the nation and the world. A natural result of these changes in the getting of food are the accompanying changes in food processing and preparation. What as recently as a few years ago took the average house-wife up to three hours to prepare for a full course, nutri-tional meal now requires only one-third the time thanks to prepackaging and "instant" cooking.

The advent of automation in the food industry leaves the propositions of hunting, fishing and foraging to the sportsman and hobbyist. In the processing plant, the food is received in "whole," fully formed condition rather than such elemental forms as carbon, hydrogen, nitrogen, oxy-gen and the like. What the factory does is clean the food, remove some of it, refine it, synthetically fortify it, polish it, bleach it, paint it, dye it, spray it, dehydrate it, reconstitute it, preserve it and perhaps package it for final distribution to the retailer and, finally, to the table. This practice of automated food processing affects the manufacture of food in the same fashion so that just as power equipment has replaced beasts of burden and freed the farmer from the backbreaking tasks of tilling, cultivating and harvesting, the mechanization process has expanded to include farm eleva-tors, farm conveyors, augers, blowers, shuttles and other contrivances. The modernized farm has become an engi-neering wonderland and the mechanized developments have greatly increased the food handling capacity of the farmer.

With the automation revolution comes a quest for increased production and greater yields which justified the financial investment of large sums for mechanization in the first place. In the process we have lost sight of the fact that food is not manufactured by mechanical innovation but by living creatures—the green plants and the animals which feed upon them. There are natural biological laws of time, growth and maturity which, when interfered with, lead to serious consequences in the organism.

In the words of Dr. William Albrecht, one of our nation's leading soil scientists, "We have succumbed to the idea that agriculture can be made an industrial procedure. But the truth is, it is a biological procedure." Biological meddling by man in the production of food leads to cellular and structural changes in the tampered with plant. Continued forced feedings of a few chemicals in high concentration fertilizers upset the balance of soil nutrition. After a period of years the natural balance of all the elements in the soils is skewed so markedly that abundantly supplied elements tend to block the availability of less abundant elements.

In his addiction to the maintenance of high yields, the farmer has resorted to an ever increasing number of insecticides, fungicides, pesticides, weed killers and countless other chemical contaminants which are dispensed in a variety of ways. Some of these chemicals are powerful poisons and have been used so extensively that traces of them are turning up in most of our grains, fruits and vegetables. Many of them are cumulative in animal tissue so that animals and humans who consume the treated crops begin depositing the poisons in their body cells for an overall effect of slow, relentless, deliberate poisoning. Human tolerance levels for many of these poisons have not yet been fully calculated but eminent chemists and researchers have concluded that the cumulative tendency necessitates the establishment of zero tolerance levels in many foods. No amount is safe. Yet, the spraying continues, and the practice takes on a hideous aspect when one realizes that some of the poisons used for crop spraying are also used in chemical warfare. Americans tend to feel secure simply because the deplorable practice is widespread and, therefore, normal. Perhaps the most alarming aspect is the fact that these sprays not only lie on the outer coverings of the crops, they penetrate and are assimilated into the edible portions of the plants through their roots so that washing, peeling and cooking cannot entirely remove the poisons.

One of the most flagrant misuses of sprays has been with the chlorinated hydrocarbon group of insecticides. The most prominent of this group is DDT, originally compounded in 1874 and given rebirth during World War II to combat an increased incidence of typhus through mosquito control. Since 1945 it had been used extensively in United States agriculture until banned in June, 1972 by the Environmental Protection Agency. In spite of efforts by environmentalists and concerned citizen groups, a drive is on to reinstate the license to distribute DDT in the United States by farmers, a group of scientists and others. The magnitude of the effect DDT can have on animals is illustrated by a story told by Leonard Wickenden, author of gardening and soil management books. To combat a spruce budworm infestation of 3,000,000 acres of land in Park County, Montana, DDT was sprayed as a pesticide. In Mr. Wickenden's words:

> *Apparently the authorities who organized the broadcasting of this immense quantity of poison over almost 500 square miles of forest were blind to everything but the armies of spruce budworms. One can only assume that they formed a mental picture of the DDT making straight for the stomachs of the budworms, considerately avoiding all other insects, bird and animal life in the forest. What happened was disastrously different.*

Mr. Wickenden concludes by quoting a Montana newspaper which tells of the mass destruction of all insect life, both good and bad, and of aquatic life in the area. The fish died after eating the poisoned insects or from lack of food because of the reduction of insect population. He then goes on to raise the question about the effects on the number of birds which may also have died after feeding on the poisoned insects.

Perhaps the main point is the fact that despite the apparent saturation of crops with insecticides, insects on plants still abound. This is because predator insects are also killed, while at the same time mutated pest strains survive because of resistance to certain pesticides. While less than one percent of insect species are considered "pests," the other 99 percent (including bees, wasps and butterflies—constituting the plant-pollinating species) are also wiped out. These innocent bystanders serve as aerators of the soil, predators of insects and scavengers of animal and plant waste, yet they too are killed.

Sir Albert Howard of Oxford University has offered his point of view that "insects and fungi are not the real cause of plant diseases but only attack unsuitable varieties and mutations of crops, pointing out the crops that are improperly nourished and so keeping our agriculture up to the mark. In other words, pests must be looked to as an integral portion of any rational system of farming. The policy of protecting crops from pests by means of sprays, powders and so forth is unscientific and unsound as, even when successful, such procedures merely preserve the unfit and obscure the real problem—how to grow healthy crops." (*World Crisis in Agriculture*, Ambassador College Press, 1974.)

To determine the effects of DDT on humans, we consult the work of W. Coda Martin, M.D., former president of the American Academy of Nutrition. Choosing patients who gave no history of occupational contact with insecticides, Dr. Martin analyzed the fat tissue of 25 human subjects for DDT contents. Evidence of DDT was found in 23 of the 25 subjects in amounts ranging from one ppm (parts per million) to 11 ppm with an average finding of 3.5 ppm. In seven of the subjects (28 percent) the amount of DDT was over 5 ppm which is extremely important to Dr. Martin since "in animal tests, 5 ppm will cause liver damage and is considered toxic." Dr. Martin's findings have been substantiated by other investigators who also found similar or slightly higher levels.

The definitive dangers of these residues have not been determined but we know toxicology well enough to say with certainty that this is not a wholesome practice. We can recall the effects of lead poisoning and other toxin produced fatalities so that I, for one, must cast my vote with those who say that chemical sprays and insecticides are dangerous and should be avoided.

The effect of automation does not end with the injurious soil and spraying practices of modern agriculture. It is also apparent in our animal husbandry techniques as evidenced in the common practice of livestock and poultry caging. Chickens are caged so that they can be force-fed, fattened more quickly and duped into laying more eggs through the use of timed electric lights. The result is an unnatural environment where, among other things, they are not allowed exercise and normal movement that accompanies the instinctual activities of scratching and pecking on the ground for food and insects. In addition, drugs and medication are administered to stimulate changes in natural physiology so that more flesh develops.

These techniques are also carried through to the raising of livestock. Hogs are given iron injections to boost weaning weights and cattle are administered antithyroid medication to cut down metabolism thereby building up fat. Castration, while in itself potentially dangerous, is routine for both livestock and poultry. Although chemical castration has been outlawed, surgical castration is practiced and results in an altered cellular structure where female hormones play a greater role in the emasculated animal's physiology. That this abnormal hormonal balance might be passed on to the consumer is speculated, but not documented sufficiently to be conclusive.

Not only has science come to the farmer in the form of mechanization and "conveyor belt" animal husbandry, but in medicine as well. The modern farmer is armed with a variety of potent medications, drugs and antibiotics which are often dispensed without caution. The obvious result is that traces of iron shots, penicillin, hormones, sulfa, antimetabolites, achromycin, expectorants, vitamin supplements and a host of other medications can be found in food. Some of these drugs and medications are necessary in the production of food for human consumption but many are fraught with the insidious danger of administration by farmers without a license or veterinarian supervision. In regular medical practice abnormal and fatal human reactions to simple antibiotic dosages have been reported and it is not unrealistic to consider that a patient may have built up an intolerance for a drug by consuming doses of drug contaminated food.

I would be unfair and irresponsible to consider all mechanical innovations on the farm as detrimental. Irrigation machinery, machinery which works the land, and machinery which eliminated brutal human labor necessary to feed an ever-expanding population are all acknowledged boons to agriculture. However, it is just as important not to overlook the danger in becoming so mechanized, scientific and technical that we force unnatural biological performances. When this occurs, nature retaliates and the price paid ridicules the economy of the process.

The biological manipulation in food production does not end on the farm, but is compounded when the food processor adds his share of pollution. Much of the remaining nutritional value in the food is removed in the processing cycle as exemplified in the refinement of sugar. The sugar grower sends 35 or more elements to the processor in the raw product; the consumer gets only three when he buys it from the grocery store. To the insult of overrefinement, the processor adds the injury of placing chemical preservatives in the remaining hollow shell we call sugar. A Congressional Committee investigating additives to food identified and listed approximately 700 chemicals in use for various purposes!

White flour, used in baking our bread, is another vivid example of modern technology "processing for non-health." The wheat milling process that results in white flour removes the following parts of the wheat germ kernel:

Bran which makes up approximately 14.5 percent of the kernel, including nucellar tissue, seed coat (tests), tube cells, cross cells, hypodermis and epidermis.

Aleurone cell layer, part of the endosperm, is separated with the bran in the milling process accounting for loss of the bulk of wheat's rich protein matter, some trace elements and useful fatty substance.

The germ is also removed in the process accounting for loss of protein, natural sugars, a considerable quantity of wheat oil and a large percentage of vitamins and minerals, especially trace elements.

The residue of this milling process is fed to animals with the obvious result that animals are on a better diet than people who eat merely the remaining endosperm that constitutes the bulk of white flour.

The implications of the milling process to those who eat white bread are interesting.

Eighty-six percent of the manganese content is removed by the milling process. Chickens and animals experimentally deprived of manganese grow improperly and often become sterile.

A large proportion of selenium is removed in the process. Rats and chickens deprived of selenium show signs of liver deterioration.

Approximately 78 percent of the zinc is removed; zinc is known to speed the healing of wounds and human dwarfs are a recognized result of severe deficiency of zinc.

Eighty-nine percent of the cobalt is removed; cobalt is known to be a key element in Vitamin B-12, is important to the maturing of red blood cells which carry iron and oxygen in all warm-blooded mammals.

Nearly half of the chromium is removed. Lack of chromium has been shown to contribute to the incidence of diabetes.

Seventy-seven percent of the Vitamin B-1 and 67 percent of the folic acid are lost; both along with other trace elements are key in the manufacture of RNA and DNA, the chemicals which pass along the genetic code having to do with the building of cells and procreation.

Eighty percent of the B-2 and 81 percent of Vitamin B are lost; both are important in mucous membrane health and resistance to Pellagra.

Seventy-two percent of B-6 is lost; Vitamin B-6 has to do with the metabolism of amino acids which are the building blocks of proteins making up most of the body.

Most of the Vitamin A is lost in the process; A is essential in the maintenance of good vision and healthy skin.

Eighty-six percent of the Vitamin E and most of the Vitamin D content are lost; E is necessary in the proper development and maintenance of cell membranes and D is important in utilization of calcium and Vitamin A.

In addition to the above more dramatic illustrations, the milling process removes 50 percent of the pantothenic acid, 76 percent of the iron, 60 percent of the calcium, 78 percent of the sodium, 77 percent of the potassium, 85 percent of the magnesium and 71 percent of the phosphorus.

In the past, the law required that any harmful chemical additives in use by food processors be detected and proven harmful by the Food and Drug Administration before the product containing them could be taken off the market. A new law now requires that chemical manufacturers first test the chemical substance on animals and submit evidence of testing results before the chemical can be approved for use in food products by processors. This is an apparent improvement over previous methods but one wonders whether individually isolated tests for toxicity and other factors are sufficient. One chemical preservative may be individually innocuous but may well be dangerous in combination with other chemicals or preservatives on the market. To our knowledge, there has been no complete study of the cumulative effects of all the chemical preservatives in our foods. It should be kept in mind that the human consumer eats a wide variety of foods over an extended period of time and the interaction of the various chemical preservatives and additives in these foods may be less than desirable.

In summary, U.S. agriculture and food processing techniques are attempting to accomplish the impossible mechanization of biology. In our driving ambition to produce more and more on less and less, we produce enormous quantities of food of dubious quality. To maintain high production rates in agriculture and animal husbandry, we resort to measures, some of which border on insanity, that include use of drugs, medications, mechanical devices, aerators and even poisons. We then over process, over refine and succeed in removing much of the already depleted nutrition in the foods. We add preservatives to reduce spoilage and to the whole procedure we imbue the concept of automation which beats from nature all she can deliver.

Our technological progress has been so rapid that we have forgotten that life consists of protoplasm—living protoplasm which is bound by the biological laws of ingestion, metabolism, growth, maturity and death. In place of the basic rules of living which have stood the test of time, we have substituted newer, more modern approaches that have not been time tested and which have all the earmarks of danger. Thus in a quest for modern, improved living standards, our future may be in jeopardy.

It is not difficult to see that is why our health is not good. What is remarkable is that our health is as good as it is. We are nutritionally deficient and, as a result, we open ourselves to attack from parasitic organisms. We submit to slow poisoning through cumulative toxins and try to get something for nothing by defrauding nature. All the problems inherent in our modern system can be eliminated with the application of sea energy in agriculture and good sense in processing.

Chapter 8
"Organic" Versus "Inorganic"

East is East and West is West
And never the twain shall meet.

—Rudyard Kipling

Once a missionary-scientist, spreading the gospel to a primitive society, overheard an elder member of the tribe tell a group of native children not to build their canoes out of trees in which a certain sacred bird had nested. These same trees, however, were considered the best source for canoes when the birds had not nested in them. When asked the reason for his warning, the elder told the missionary-scientist that the Bird God "destroys canoes made from trees from which the sacred bird's home bad been taken by filling the trees with evil spirits to plague the human who violated this sacred decree."

The missionary-scientist scoffed at this ancient belief and set out to prove to the tribe that evil spirits did not possess these nested trees. He built himself a canoe from the taboo tree and, while all the tribe gathered on the beach to watch, set out into the bay. When he had paddled the craft just beyond a nearby reef, it sank and the missionary drowned. The primitives returned to the village and continued worshiping their gods.

We tell this story because it illustrates a simple natural truth, to wit: some people can be right in practice but wrong in theory while others can be right in theory and wrong in practice. In our example, the natives practiced accurately while the theory behind their practice was scientifically inaccurate. The missionary-scientist, on the other hand, was right in not accepting the evil-spirits doctrine, but failure to thoroughly investigate the facts cost him his life. The truth of the matter was that the sacred birds had nested close to a food supply and the wood of the trees in which they nested was filled with wood-boring worms.

Somewhat analogous to our story is the great debate in progress in the United States today with regard to practice and theory of growing plants for food. The debate has evolved because the principles have encamped themselves in one or the other of two opposed agricultural theories: Organic versus Inorganic, or chemical, farming.

Biologically speaking, the organic, by definition, pertains to living tissue or protoplasm. Therefore, it refers to anything derived from or exhibiting character peculiar to living organisms. Organic has a chemical definition as well. Here organic pertains to that branch of chemistry dealing with compounds of carbons and, thus, is a study of the chemistry of carbon.

Inorganic is defined as "not organic" and is composed of matter other than vegetable or animal. Alternatively, inorganic matter is matter that is inanimate and lacks possession of characteristics peculiar to living tissue. Inorganic Chemistry then is the branch of chemistry dealing with all substances except those referred to as organic.

Now, to the debate.

In essence, the proponents of the organic growing method hold the theory that plants are composed of nutrition which is supplied by decomposed living matter contained in the soil. Soil enriched with forms of decaying life such as manure, mulch, sewage and sludge are the best source of plant nutrition. They argue that this is the "natural" method of growing crops, and therefore, it is a superior method. At the same time, they contend that plants grown with chemicals and artificial fertilizers are inferior, lacking in nutritional elements and, in many ways, detrimental to the health of both the plants and the animals that eat them. Additionally, they blame many of the nation's physical ills on the "poisoning" of the soil by unwise use of chemical fertilizers on the nation's farms. In practice the organic farmers grow crops of excellent taste, flavor and high nutritional value.

Advocates of inorganic or chemical farming, on the other hand, contend that they are able to grow excellent crops by adding to the soil various minerals and chemicals which the plant utilizes more efficiently than organic fertilizers. They report carefully controlled studies of crops grown on soils to which chemicals have been added that show no difference from crops grown on organically fertilized soils. The use of chemical commercial fertilizers is also a much quicker, more feasible way to grow crops and bring them to maturity. These inorganic proponents cite high crop yields and accompanying increases in average human life expectancy in the United States as proof that their method is best and that soils are not being poisoned in the process. Instead the inorganic advocates contend that what primarily affects the nutritional composition of a food is the genetic makeup of the seed, not the soil fertility. They consider the organic farmers to be "faddists" and accuse them of dabbling in quackery because of their belief that illnesses are the result of soil depletion and malnutrition. Moreover, the organic advocates have not scientifically demonstrated the superiority of the organic method.

With this brief background describing the main points of contention, let us examine the facts.

It is true that organic farming imparts some very desirable properties to the soil; that soil texture is often improved; that the nutrients in an organically fertilized soil are good; that plants grown with organic fertilizers have good nutritional value and perhaps superior taste when compared with those grown under some forms of chemical fertilizers; that the building of microscopic life in the soil is beneficial to the crops grown; that aeration and supply of oxygen to the plant roots is good in organic farming.

However, it is not true that plants need organic matter in order to grow into healthy living organisms. It is not true that chemical fertilizers "poison" crops or the soil per se.

What the organic farmers overlook is the fact that almost without exception *plants utilize elements for their nutrition only if those elements in the final analysis are in an inorganic state before absorption.* Regardless of the form of the fertilizers when they are placed in the ground, they must first be converted to an inorganic state before the plant takes them in! This is the essence of the primary difference between plant and animal life on this planet. Plants take in elements in the inorganic form and convert them to an organic form. In an opposite manner, animals must have elements in the organic form in order to carry out their metabolism. Thus, all animal food is organic with the single exception of common table salt, a compound which is currently under considerable medical scrutiny with regard to its role in chronic disease and toxicity implications.

At first water appears to be an exception to the organic requirements of animals since it is inorganic. However, all water must be "bound" organically with protein, etc., before entering the bloodstream. Nearly half the human torso is devoted to this process and it is interesting to note that the inner surfaces of the lungs, which are exposed to oxygen, would stretch over an area as large as a tennis court.

Contingent on the discussion of organic and inorganic elemental forms is the distinction between infection and decay that is applicable to the successful practice of medicine, both in diagnosis and treatment. Those bacteria which are plants do not ordinarily utilize organic substances upon which they may be living. Instead the mode of digestion for bacteria is characterized by first breaking down the organic elements into an inorganic form. To accomplish this, the bacteria must have free water with food substances dissolved in it. A product of the bacterium's metabolism then breaks down the organic compounds into simpler substances that can be ingested. If the bacterium is residing and working on a living organism in the form of plant or animal tissue, the breakdown of organic matter is known as infection. If the bacterium's object is dead animal or plant tissue, the breakdown of organic matter is called decay. This distinction between infection and decay is sometimes overlooked by members of the healing professions and their practice is poorer as a result.

When evaluated under botanical facts, the organic hypothesis which demands that all fertilizer be organic in order to be beneficial is rendered invalid. The criticism that chemical farming poisons the soil also breaks down under this precise analysis. While overly harsh, this criticism is not without foundation although chemical fertilizers do not "poison" the soil or crops per se. Instead, the unwise manner in which they are disproportionately applied to the soil can and does upset the physical and chemical functions of the plant and can produce blocking of nutrients or an imbalance of elements.

Let us now evaluate the points used by the inorganic proponents to substantiate their arguments. It is true that plants can grow through their entire life cycle without organic matter; that organic matter has no magical properties; that yields are very high for chemically fertilized farms; that the average life expectancy in the United States is statistically higher than in most other countries; that under chemical fertilization our farm production is the highest it has even been.

But, it is not true that plants grown under the present methods using chemical fertilizers are as vital and complete nutritionally as the claims would have us believe. Furthermore, it is absurd to say that the genetic makeup of the plant's seed is more important to nutritional composition than the plant's food in the soil. It is not true that based on average life expectancy, American health is good. But most important, it is not true that supplying a plant with its five or six major elements in the form of a fertilizer is sufficient to produce good and healthy plants. Plants require a great deal more than five or six elements.

The debate between "organic" and "inorganic" farmers is then reduced to the point of our chapter. The "organic" farmers are right in practice but they adhere to a clouded theory. Conversely, the "inorganic" farmers subscribe to correct chemical theory, but they are woefully inadequate in their biological practices. An accurate overview would serve to combine the inorganic theory with organic practice to provide all the necessary elements of nutrition in proper quantities and optimum balance. If we stand back and observe the facts in the debate, the overall truth may present itself more clearly.

First, organic farmers are producing crops in which the primary concern appears to be quality and the secondary concern quantity. Conversely, chemical farmers seem primarily concerned with producing crops in great quantity with secondary concern for quality. Secondly, if plants can be carried through their full life cycle without organic fertilizer and even without soil as in hydroponics, how can we offer explanation in a coherent and accurate philosophy of botany which refutes the organic growers' precepts, yet recognizes their excellent results? Third, if the advocates of chemical fertilizers have accurate botanical theory, what is incorrect about their practice?

The answers are relatively simple. The organic farmers have high quality crops because with their fertilizing practices they put a greater number of elements back into the soil in proportions which once made up living substance. Before the plant can utilize organic substances, however, the substances must be first reduced to their inorganic form. Physical agents such as freezing and thawing as well as the macroscopic and microscopic life in the soil help to bring about this change.

As earlier mentioned, the organic farmer does not feed his plants "organic" food in terms of strict chemical analysis since the organic must first be reduced to inorganic form to be assimilated by the plant. It bears repeating then that *all plant food is inorganic, and all fertilizers are, therefore, inorganic.*

In dealing with the second point, we must agree that chemical farmers enjoy high farm yields and crops which appear to be of good quality, but they fail to recognize the inadequacies of their fertilizers. Major nutritional elements are put into the soil and, with heavy nitrogen stimulation, plants are grown which are observed to be every bit as good nutritionally as those grown with organic fertilizers.

Under discrete spectrographic analysis, this is not the case and chemical farmers overlook the fact that they are putting an incomplete plant diet on the soil. Such practice upsets the balance of nutritive elements by stressing the importance of major elements and, in time, leads to abnormal soil conditions. Plants then may take up abnormal proportions of other elements to make up for the deficiencies in trace elements, crops begin to lose their nutritional value.

Humans, having lost their keen senses of taste and smell in the natural course of evolution, are unable to appreciate these subtle nutritional losses as readily as the lower animals. In scientific literature, numerous examples are cited of animals forsaking beautiful looking crops for scrub grass, weeds, nettles and the like, and for less luxuriant growths of plants in soil where trace elements have been added. Humans are forced to rely upon laboratory analysis to determine the absence or presence of nutrients in their food.

Because the chemical farmers fail in fertilizing practices to supply complete plant nutrition, the argument is, in my mind, resolved. I cast my vote for the practices of the organic farmer and accept the plant physiology theory of the inorganic farmers. Thus, when asked for a prescriptive plan of action, I suggest the practices of the organic farmer, while recognizing their hypothetical shortcomings. For the inorganic advocates, I suggest using a complete chemical fertilizer in which the trace elements are included in balanced, life-supporting proportions. And finally, I say do not knock the "organic" approach because in practice it works; in both the scientific and natural worlds there are many examples of things that work for which we have no scientific understanding. One day, it is hoped, we will understand because if East is East and West is West, the twain of true thought and true practice must meet even though they travel in opposite directions. After all, the world is round.

Afterword

Dr. Maynard Murray passed from the scene in 1983. Except for a few disciples, his vision and work became shelved, the only explanation and analysis lay out of print. The art and science of sea solids agriculture was recovered when it became front-page fare in *Acres U.S.A.*, and from this re-exposure developed an enlivened interest and a new demand for the re-issue of this book as a solid entry in the growing shelves on biologically-correct agriculture.

Dr. Murray's demurrer on the word *organic* should not prove to be troublesome. Every word in the English language has several meanings, it being the function of literacy to discern the meaning of a term in the context of its usage. *Organic* as used in agriculture generally means *naturally grown*. Chemists have given the term their own name, hence things like chemicals of organic synthesis, as in organic chemistry. Just the same, Dr. Murray's manuscript, as assisted by Tom Valentine, stands as first written, a monument to the inquisitive nature of the true scientist.

After Maynard Murray passed from the scene, it fell to a Nebraska farmer named Donald Jansen to pick up the mantle for sea solids agriculture. The Mennonite colony near

Ogallala, Nebraska, which provided the ethical and farm training for Jansen was half a continent away from Murray's experiments. If anything, an unseen hand seemed to guide Jansen through Northwestern University, several theological seminaries, the ministry, and an Ohio University teaching post. The Nebraska home farm had grown from one section to 15,000 acres, replete with Angus cattle, 5,000 acres of wheat, 50 head of buffalo—an industrial farm that so pressured a brother, he killed himself.

In 1978 the family and the bank required Don Jansen to come home and take over. He left university tenure behind. A farm of such scope was "massive insanity"—Jansen's phrase—for which reason the judgment was made to sell out.

Circa 1949, with the establishment of Poison Control Centers, public policy decreed that henceforth agriculture in the United States would be toxic agriculture. Jansen's brother had bought into all the university technology. This meant poisons, sprays, multiple sclerosis, and a painful death. Moreover, toxic technology seemed to have delivered cancer to Jansen's aging father. In order to save his father from certain heart failure, Jansen turned to chelation, and while attending his father, he took chelation himself. While the drip bottle emptied, another patient presented Jansen a copy of this book in its first edition.

Jansen contacted Murray, then ordered a semi of sea solids, origin, Baja, Mexico. The first innoculation was on pasture before the spring rains. The resident buffalo mowed on that pasture as never before.

The next experiment was on wheat, a quarter of a section. The results were incredible, as mentioned in Murray's text. The wheat came up gangbuster style. It covered the sandy hills, the valley—it was gorgeous wheat. The mineral-starved soil responded as though a magic wand had been passed over it. Jansen's Mennonite neighbors looked, but they refused to believe what they saw or were told.

Jansen did close out that Nebraska farm. Instead of pursuing the elusive nothing called industrial production agriculture, he took an interest in Murray's Florida acres. Murray was near retirement and clearly not a farmer, but a laboratory scientist and a whole-foods afficionado. Murray was also a doctor at a mental institute in Fort Myers. Jansen spent one year with Murray before that genius of a man passed away.

Since that time, 1983, Jansen has grown almost every crop with seawater. Each experiment has caused nature to reveal herself. One serendipitous finding was that peanuts could be grown pre-salted.

Those two decades of work have proved that poison agriculture is merely a monument to stupidity. Hydroponics, rather than inviting disease, serves up absolutely clean production. Tomatoes grown in gravel with Murray's solutions feeding the plants told Jansen they could beat soil-raised plants by a week to 10 days.

No one has fed farm plants or people 92 elements all their life. This one fact leaves a vacuum, one that research will take several lifetimes to unravel. Dr. Maynard Murray believed the ocean was balanced, that God made it that way. It is no mistake that sodium is first and 91 other elements are lesser.

Different plants take different solutions, both Murray and Jansen hold. Taste comes first, then it is backed by analysis. Shelf-life for vegetables and fruit loaded with minerals has proved to be outstanding. Total nutrition seems to equal anti-fungal, anti-bacterial properties. The Murray findings and the Jansen refinements became a business that shipped all over the United States.

In 1988 Jansen bought a hydroponic farm in Florida. The signal word is total nutrition. When total nutrition is taken up systemically by the plant, it puts all of its energy into the plant and its fruit and seed.

The road has been rocky for the Murray-Jansen development. In spite of quality, lavish production, real support was not forthcoming. Jansen backed off until recently—but now the technology for mass production of sprouts for juicing, and suitable preservation for juice for distribution have revealed themselves.

In tune with the rapid acceptance of clean farming and an even greater concern about nutrition by the population in general, a paradigm shift in potential has arrived. The hour has arrived anew for ocean-grown production to take off as envisioned by Maynard Murray in the first place. The next step will be to take sea solids agriculture into the fields, the greenhouses, the hydroponic units. After all, there is a relationship between health and nutrition.

—*Charles Walters*

Index

cancer, 4; bowel, 78; breast, 50, 73; colon, 79; in trout, 30
carbon, 28
carbon atom, 74
carbon cycle, 4
carbon dioxide, 15
carrots, 26; experimental plot, 44-45
catalysts, 74
cattle, 8; fed diet grown with sea solids, 52
cells, 1; function, 16; nutrition of, 16; red blood, 16; compensation for nutritional shortfalls, 69
chelation, 102
chemical contaminants, in food, 83-84
chemical denudation, of land masses, 29-30
Chicago Medical Society, 8
chickens, 8; feeding experiments, 48-49, 72-73
chlorinated hydrocarbon, 84
chlorine, 25, 38
chlorophyl, 2
chlorosis, 25
cholera, 12
cholesterol, 52
chromium, 89
cobalt, 25, 88
cochlea, 18
colloid state, 30-31
copper, 16, 25, 38; deficiency, 25
corn, 8, 26, 36, 52; field experiments, 45-48, 52-54
corn blight, 27
cows, fed diet grown with sea solids, 52
Crops and Soil Magazine, 33
crystalloid state, 30-31
DDT, 84-86
decay, 97

dental disease, 12
depletion, of soil, 24
dermatosis, 12
diet, American, 6
diphtheria, 12, 18
disease, 68; chronic, 69; resistance to, 35-36, 51-52; geographic variation, 78-79
DNA, 89
drugs, in food chain, 87
economics, 75
electricity, in cells, 1
elements, for hydroponic growth, 38; balance in sea water, 70
Elmhurst College, 49
endocarditis, 18
Environmental Protection Agency, 84
epidermophytosis, 18
erosion, 21, 70
exanthema, 25
eye disease, in rats, 73
famine, 68
feeding, experiments, 50-55, 72-73; selective, 18
feldspar, 78
fertilizer, 69; importance of balance, 74; inorganic state of, 96; and incomplete plant nutrition, 99
field tests, summary, 54-55
Finland, Maxwell, 20
Flewelling, Ralph Tyler, 67
flour, 88-89
fluorine, 25, 78
folic acid, 89
food, gathering, 81-82; preparation of, 82-83; processing of, 82-83; drug contamination of, 87; nutrition loss, 87-89
Food and Drug Administration, 89

Parke-Davis and Company, 77
Pasteur, Louis, 20
Payne, Eugene H., 77-78
peaches, 26; experimental plot, 40-41
pears, 26
peas, 26
pesticides, 83
phosphorus, 3, 4, 24, 31, 38, 89
photographs, of research, 58-65
photosynthesis, 15
pig, 8; feeding experiment, 49; as experimental animal, 55
pineapple, 26
pituitary, 16
plants, nutrition of, 3
pneumococci, 20
Poison Control Centers, 102
pollution, of food, 81-91
Pories, Walter J., 34
potassium, 4, 24, 31, 38, 74, 89
potassium nitrate, 70
potatoes, 26
preservatives, chemical, 90
production, methods of, 27; high yields, 83-84
protoplasm, 90
rabbit, experiments, 52
Rabelais, François, 11
rain, 21
rat, experiments, 51-52, 73
Ray Heine and Sons, 45, 50
recycled seawater, 68
rheumatism, 13
rice, disease, 26-27
RNA, 89
roots, and salinity, 7
Rosette, 25
Rosner, Lawrence, 43, 44
Rutgers University, 25
Science News, 32
sea, biological activity of, 68-69
sea energy agriculture, 5
sea energy, technology, 37-55

sea plants, 32
sea salt technology, 75
sea salts, 3
sea solids agriculture, 101, as fertilizer, 2, 3, 4, 36, 45-55, 71-73; in hydroponic solution, 39; and disease resistance, 51-52
sea water, 30
seed, variety, 27
selenium, 88
Shakespeare, William, 20
silicon, 38
Simkovitch, V. G., 21
smallpox, 12, 80
Smiles, Samuel, 29
sodium, 25, 38, 89
sodium chloride, 7, 30, 33, 39
sodium nitrate, 70
soil, 21-28; exhaustion of, 26; abnormal conditions, 99
soybeans, 8, 26
St. Mary's Hospital, 78
steers, 54
steroids, 79
Stokesberry, Paul W., 45
Strain, William H., 34
streptococci, 20
Streptococcus viridans, 18
streptomycin, 18
Stritch School of Medicine, 50
sugar, 87
sulfur, 16, 24, 38,
Sunland Center, 9
tetanus, 12
The M.D., 78
This Week Magazine, 77
thymus gland, 51-52
Tobacco Mosaic Virus, 42
tomatoes, 26, 74; hydroponic experiment, 41-42, 42-44
top soil, 21, 70
trace elements, 25, 32, 74; in nature, 29-36; toxic levels